Living off the Land

My Cornish Smallholding Dream

Lorraine Turnbull

Copyright

Fat Sheep Press, Le Bois Vert, Lescarpedie, 24220 Meyrals, France.

British Library Cataloguing in Publication Data

A CIP record for this book is available from the British Library

ISBN 978-1-9163890-2-1

Testimonials

"Well, it is FANTASTIC. I've laughed and shed some tears. It's one of those books you just have to keep reading! It's been in the garden, the kitchen, the bedroom and even the bathroom. Brilliant!"
Gail Parsons

"A truly amazing story and very readable."
Madeleine Bath

For other testimonials please see the entry for the book on the Amazon website.

Other books written by Lorraine Turnbull

The Sustainable Smallholders' Handbook (2019)
How to Live the Good Life in France (2020)

Connect with Lorraine:
Facebook page
https://www.facebook.com/LorraineTurnbullAuthor

Twitter Lorraine Turnbull
https://www.twitter.com/LorraineAuthor

Instagram Lorraine Turnbull
https://www.instagram.com/lorraineauthor/

Acknowledgements

Thank you to all my readers without whom of course, there would be no more books! Special thanks to my fabulous reader, editor and friend Rod Haselden-Nicholls and to my good friends Joanne Ainscough and Cheryl Arvidson-Keating.

Finally, as always I am forever grateful to my patient husband John and my children who have allowed me to live the life I lead and encouraged and supported me throughout.

About the author

Lorraine Turnbull wanted to be a farmer since she was five years old. After running a successful gardening business in Glasgow she uprooted herself and her family and moved to a run-down bungalow with an acre of land and an Agricultural Occupancy Condition in Cornwall. She retrained as a Further Education teacher and taught horticulture at both adult and further education level, whilst running a one acre smallholding. After working as a Skills Co-ordinator for the Rural Business School, she began commercial cider making in 2010.

In 2014 she was recognised for her contribution to sustainable living by winning the Cornwall Sustainability Awards Best Individual category. Her first book, *The Sustainable Smallholders' Handbook* was published in 2019.

Lorraine now lives in the Dordogne with her husband, cocker spaniels and sheep. She is currently writing her fourth book.

Introduction

Do you ever have a recurring dream? No, not the one where you're stark naked in a crowd of people. A wonderful dream where you are finally content; at peace and happy because you have finally achieved that *thing* that you have been searching for. You may have been working towards it for many years; but it's always been just out of reach, elusive, tangible but you can't quite get there? If you've had a dream like this, then you might be able to understand my recurring dream.

It's sunny. I'm working in the orchard outside a cottage with white-washed walls. The cottage is surrounded by immaculate, gently-rolling green fields full of freshly-laundered fluffy sheep, and the yard filled with the clucking of chickens. The air is scented with the tang of wood smoke and honeysuckle and I *know*; even though it's a dream, that I'm insanely and supremely happy and content.

I've had that dream ever since I was five years old. I used to tell Alex Wotherspoon (my then boyfriend) that we would get married, live on the Isle of Lewis and grow potatoes for a living. None of that ever happened, thank God, and I'm sure he's eternally grateful; in fact the thought of any of this coming true fills me with horror. But, I have to tell you that to the secret admiration of some of my cousins and friends, despair of my mother and frequent embarrassment of my children, I managed to live my dream.

This is the real story about how I dragged my new husband

and children from Glasgow to rural Cornwall, started a smallholding, ran a successful cider making business and finally overcame the crushing obstacle of an Agricultural Occupancy Condition. It's been a period in my life crammed full of failures and successes; of making new friends, living in a stunningly beautiful and historical part of Britain and having some of the most memorable experiences of my life. It's also been a journey of self discovery, fulfilment and of heartbreak and doubt. My children are adults now, pursuing their own very different careers, but they tell me they have wonderful memories of living in Cornwall.

Living on a smallholding can be exhilarating, exasperating, expensive and exhausting. Just to "live simply" requires you to simplify your life substantially, reduce your wants and work with nature. It's not a choice for the faint-hearted; you really have to want to live this lifestyle. The modern practice of *permaculture* is a broad re-working of this idea, and the Covid-19 pandemic has made the world respect all farmers, growers and food producers, big and small. This book has no pretensions to being a model for idealised small scale farming; it is simply a fond, very personal recollection about my journey towards living sustainably on a smallholding in Cornwall. Some of the names have been changed to protect anonymity, but this is a true story. For readers who fancy starting their own smallholding journey, you really need to read my first book, *The Sustainable Smallholders' Handbook*. It's a wonderful life.

1.
A Desperate Thirst

In my head I've been a farmer since I was five years old. I started with a vast amount of *Britain's* toy farm animals on a large painted board with green fields and blue ponds. I had flocks of sheep, a small herd of cows (these were more expensive to buy with my very meagre weekly pocket money), some poultry and a few 'human' helpers; a plastic girl with outstretched arm and a farmer who got a bit chewed up by my careless mother's vacuum cleaner.

I would spend hours and hours daily, moving the animals around on a green painted board, projecting my personality into the small plastic figure of the girl with the ponytail and outstretched arm feeding all the animals in turn. The fact that I can still vividly remember all the pieces even now will tell you how much I cherished them and my little childish farming fantasy.

Because we lived in a run-down suburb near Paisley where social deprivation was the "norm" and school-age pregnancy more and more common, my parents, having a car, took us on endless trips to the countryside. Strawberry picking, fishing, watching pheasants whilst collecting brambles (blackberries) and walking ensured we never made friends with the wrong sort of people. When I look back, I only have vague recollection of people from that time in my life. I can only vaguely remember the 1960's; not because of drugs but because my mother controlled every single thing in life. So, we went to church and Sunday School because my mother was living in a Doris Day fantasy life.

Determined to raise our socio-economic status, she forced my dad to accept a job working abroad in the oil industry, and secured a crippling mortgage on a newly-built semi-detached house in a fashionable south-side suburb of Glasgow. We inherited a large rectangular bare earth plot from the builders; which by no stretch of the imagination could you call a garden, and my mother had no idea what to do with it. However, I was delighted; packed away my farm animals and discovered gardening.

Trips to the local library ensured I had enough reference books on design and plants and my subsequent success creating a large lawn encouraged her to allow me to begin propagating plants and shrubs. My mother had no interest in gardening. She merely wanted it to look pretty and be somewhere to sit out on a nice day. A builder was found to construct a small patio area joining the house to the garden, edged with small narrow beds soon filled with flowers and shrubs that I had been busy propagating. I dug and planted a vegetable patch at the rear of the garden which produced beans, peas, raspberries and blackcurrants and a few leeks. An apple tree rescued from the local garden centre's Cemetery Corner survived and once planted out was so grateful for the rescue that it went on to produce apples every autumn. My pocket money was meagre but I managed to save a little for the odd horticultural treat, when I wasn't rescuing animals.

My mother; for the most time, tolerated my appearing with small animals who needed re-homing. First, Snowball the hamster, then Snowy the rabbit. Snowball sadly escaped and disappeared forever somewhere in the piping behind the kitchen cupboards. Snowy turned out to be a psychotic aggressive demon and would bite and kick at every attempt to pet or cuddle and at cleaning out time. When I 'rescued' two mice from the local church fete's racing mice stall she put her foot down. They had to live outside and very shortly after being brought home, they somehow managed to 'escape' their cage in the garage. I found an old round plastic goldfish bowl and after a day spent at our local swampy pond caught a few newts to keep in it. In those days, as a small child, I had no idea of the rules regarding newts or any wild creatures and was just satisfying my curiosity and love for creatures. These also soon disappeared and I was strictly warned No. More. Pets.

Horse riding lessons were provided as this was considered a genteel pursuit for a young lady and would provide the animal interaction I craved whilst mixing with other girls from the 'right sort' of families. I spent every non-school day at the stables, mucking out and improving my riding, and generally looking and smelling like an agricultural worker. Lessons were then stopped, as they had failed, in my mother's view to turn me into a lady. I begged for a horse of my own and was instead eventually palmed off with a Cocker Spaniel puppy. Scamp wasn't even the colour I would have chosen, but had been picked by my mother to match the furniture and carpet. Nevertheless, we became inseparable, and disappeared for hours into the fields and woods nearby.

By the time I reached fourteen, I had a weekly subscription to *Farmers Weekly* much to my mother's horror and despair. She longed for me to magically turn into a lady overnight and to become a hotel receptionist or typist, and then to get married and present her with grandchildren. I wanted to go to agricultural college to explore the fascinating world of sheep fertility and how to get three lamb crops every two years. Even then we were poles apart, and looking back, I must have been a total mystery as well as a disappointment to her.

Goodness knows where all this affinity for the land came from. My father was from Govan; a hard and deprived ship building area located on the River Clyde. After leaving the Navy at the end of WW2, he trained as an engineer, and worked in the Middle East in the oil industry, which is how he met my mother; at a party in Baghdad, Iraq in the 1950's. My mother was Armenian, and in her youth, quite a beauty. Her own story is fascinating. Her grandfather was an Armenian carpet weaver in a small town on the old silk route in Western Turkey. He was killed when the Turks massacred the Armenians in the genocide in the early twentieth century. The survivors, including his son (her father) who was aged about seventeen, were marched South East across the Syrian and Iraqi deserts. The death marches, as they became known, starved many, and brutal robberies, rapes and murders were commonplace. Many thousands died along the way, although the Bedouin and various international charities tried to help. Her father walked with others to Baghdad, where he met his wife; a fellow Armenian refugee and they married and started a family. Sadly, the wife died very young, leaving the husband with four young children to look after. He remarried and went on to have another four sons.

How much this influenced her choice to marry I'm unsure. She could have had her pick of men, but chose to marry my dad; leave a tight-knit community and family in a country now under British control, to live the high life in the 1950's mainly-American-foreign -workers compound, with servants and horses and parties, before finally returning to post war, smog-filled Glasgow. She must have really loved him. Or have been very desperate to leave post-war Baghdad.

She was the sole parent in a time before single parents were accepted as normal, because my father worked abroad in the oil industry and we only saw him for a couple of weeks twice a year. She ran the house, made the rules and attempted to mould me into what she wanted, by a process of guilt and bribery. They say opposites attract, but my mother and I just could not agree on anything and the only thing we actually had in common was our stubbornness. Even at this early age, the thirst for a rural lifestyle was stronger than my mother's need to imprint her wants and needs on me. Not for me high heels and make-up, but Saturday visits to Smith's bookshop in Glasgow, where I bought *The Production and Management of Sheep* by Derek H. Goodwin (Hutchinson, 1971); which you can still buy if you hunt for it, and a damn fine read it is. My father; when he returned home for short periods between working abroad in the oil industry, despaired of my lack of academic brilliance. I remember the only parent teacher meeting he attended, shortly before I sat my Ordinary Grade exams. The head teacher; a disgrace to her profession in my memory, pronounced that I was a failure; but could probably get a job on the line at the local pickle factory. I never forgot that. Nor his failure to champion my successes.

Although an early and accomplished reader, my incompetence in arithmetic was legendary. Despite my best efforts, I managed to fail my Arithmetic Ordinary Grade (the equivalent to a GCSE) in secondary school on no less than 3 occasions. It didn't matter to Dad that I was getting straight A's in everything else, or even that I was very good at managing money, that damn arithmetic certificate was the Holy Grail; and for me, just as elusive. My grades in Biology, Chemistry and Geography were good and my teachers encouraged me to progress, and my garden; *my* garden, was filled with flowers and shrubs all grown either from seeds or cuttings I had cultivated. When I needed peace and to feel valued, I escaped to my garden and my plants. Here at least, I knew I was successful at something.

Anyway; as if I wasn't disappointing enough, I finally managed to firmly throw the shit at the fan at the age of seventeen when I secretly applied for a place on an agricultural course at The Robert Gordon Institute in Aberdeen. When the letter arrived, offering me an interview, my mother intercepted and opened it and was hysterical, demanding to know how I could "*do this to them*". My father was quietly furious but agreed to go with me to Aberdeen for the interview. My anger at having my mail opened was slightly mollified by the fact that I'd secured an interview.

I endured a week of my mother making loud, distraught phone calls to her brother; crying down the phone that I "... wanted to become a *...farm labourer!*" So eventually the day came and Dad and I boarded the train into Glasgow and the connecting train to Aberdeen. It was a long journey and the atmosphere was tense. At the college, I was interviewed and offered a place unconditionally. They were more interested in attracting female students with aptitude than academic brilliance and I was confident and very interested. I was overjoyed, but that feeling was to be quickly crushed. On the return train journey I was informed that "under NO circumstances" would I be accepting the place, that my mother had registered me on a hotel receptionist course (you can imagine my eye-rolling at this revelation), and then he trotted out the 'whilst you are living under my roof' speech.

Even looking back now, forty years later, I can still remember every detail of the pain and anger I felt at that time. Far from being supportive and delighted that I had overcome my supposed lack of direction and perceived intellectual failings, my mother was distraught that people would look down on the family because I was studying farming, and my father thought that, as a female student living in digs in Aberdeen, I would "naturally" be prey to the oilmen working in that city, and "bring shame on the family". I think what hurt even more was the fact that he had encouraged my hope by allowing the journey to Aberdeen, to the interview, and only when I had secured a place did he deliver the killer blow. If this happened today I'm sure it would be construed as mental abuse.

So with my dream kicked into touch, personal self-esteem in tatters, and a whole new understanding of my worth to my parents, I cried for months and never forgave either of them. I lasted for a few months on the hotel receptionist course, bored stiff and thoroughly miserable, before giving it up. Years later, my mother denied that the 'interview' incident ever took place. I left home as soon as I could, at the age of 23 to live in digs in Edinburgh and attend a college course to qualify as a library assistant. The course was full-time and I worked evenings stacking supermarket shelves and then nights at a bar to make ends meet. I buried my dream and worked as much as I could, securing a full time job at the National Museum of Scotland, and my first mortgage on a one bed-roomed flat in Leith.

I worked hard and renovated. I could afford to pay the mortgage, walked to and from work and occasionally ate. After a couple of years I sold my tiny flat, bought a better one and renovated that. I changed my job from Library Assistant to become a Police Officer in Strathclyde Police; and doubled my salary overnight. This allowed me to buy a small terraced house with a derelict back garden to renovate. I never really saw myself as a career Police Officer, although I was more than capable. The job was interesting and paid well, and frankly this was my primary concern, not career progression.

My mother saw me on the six o'clock news on television one night, at the scene of a particularly juicy murder (this was Glasgow remember) and the next morning phoned the area Divisional Commander demanding that I was moved to a safer area. I was summoned to his office that afternoon and loudly reprimanded. I was under no illusion that I would have to ensure this wasn't repeated.

Working as a female police officer was exciting, dangerous and never ever boring. I joined the same day as ten other girls, one of whom remains a good friend even to this day. Although we were generally expected to do the same duties as the male officers, there were certain incidents that became the preserve of female officers when possible, including sexual assaults and anything involving children. One of my saddest memories was attending a late evening call at a house where a cot death had been reported. The mother had refused to allow the ambulance crew and the doctor to remove the body of her baby. I sat and cried with her in the bedroom as she cuddled and held her baby daughter close. Finally; some time later, she handed the tiny bundle to me and I carried her out to the waiting ambulance. Word gets around fast in these situations and the divisional radios that night were subdued without the usual banter and chatter. My male colleague, himself a new father spent the remainder of our nightshift crying.

*****

When I finally buckled under relentless maternal pressure to get married I had just completed renovating house number four; each previous house having made me a profit and allowing me to buy better and reduce my mortgage. My husband-to-be was not a Police Officer, but was in industry. My mother was delighted. He was a manager, had prospects and good-looking. What was I waiting for? Despite both mothers keen for us both to marry, I insisted on having a long engagement, and eventually we embarked on married life. The stress of the job with unsocial hours and being held back due to unavoidable incidents, together with the new dynamics of married life, led to pressure from my husband and my mother to leave the police. Soon after this, my husband was made redundant, I discovered I was pregnant and we were forced to move to Lancashire for his new appointment, where I found myself trapped in a suburban estate with a baby; thoroughly bored with no sense of worth. I had no family or friends close to me, the house situated on an exclusive small estate and the garden perfectly designed and manicured and needing nothing more than the grass cutting.

My mother encouraged me to attend mother and baby groups. I went twice, but had nothing in common with mothers who dressed to match their baby's outfit and admitted to changing their baby's outfit two or three times a day; not because of nappy leaks or regurgitated food but *for fun*. My husband left for work before 8am and returned at 9pm and didn't feel the need to have any parental involvement. Conversation between us stalled. After all; when you've seen no-one, gone nowhere and done nothing, it's a pretty one-sided conversation. After many arguments and protestations of work changes, a second baby arrived, and things just continued as before.

*****

For a few years I led a Shirley Valentine life. With repetitive children's TV in the background and an endless and lone cycle of changing, feeding and placating children I soon resented the marriage. He arrived home when the children were asleep; ate, handed me his dirty laundry and settled himself in front of the TV, whilst I craved mental stimulation. I decided to pick up my Open University studies again and enrolled on a new module, but being there to cover a two-hour tutorial slot once a fortnight was too much for him to manage. I arranged a neighbour's teenage daughter to baby-sit, but then he changed his mind and grudgingly agreed to come home early to allow me to attend the sessions. This ploy lasted a few weeks then "emergencies" at work meant he was 'unavoidably' detained and I began to miss more and more tutorials and I was eventually forced to abandon the module. Resentment increased, the marriage foundered and I wondered how in the space of a few years I had gone from an independent successful woman to a dependant drudge.

I took the children for a short weekend visit to family in Scotland and whilst out for a walk with my brother, I tearfully explained my unhappiness.

'What happened to you?' he said. 'You don't exist as a person any more!'

And it was true. I had ceased to have a personality, and had become a mindless automaton who carried out the same repetitive tasks daily, and had no reward, no recognition and no future. There was no point trying to explain to my mother that, as a couple, my husband and I had nothing in common and that I wanted more out of life than cooking meals and polishing shoes for someone who spent ten hours a day at work and invented obstacles to prevent me completing my Open University degree. According to her, taking 'the baby' to mothers' groups should be fulfilling enough, and was 'more than she had in her day'.

It was less than a month later when, on a typical night, my darling husband handed me a pair of shoes at ten o'clock asking for them to be polished for the next morning, that something just snapped. I had a sudden mental image of me plunging a bread knife repeatedly into his back and stuffing his bloody shoes in his gaping dead mouth. Delightful and horrific at the same time, I realised the time had come to withdraw myself from this damaging "relationship" and rejoin the living.

It's a known truth that anyone leaving a long-term relationship will quickly find out that basically, You're On Your Own (which is probably the main reason that most people stay together and remain miserable). Suddenly, you realise your friends feel awkward because they don't want to take sides; your family are apprehensive for a variety of reasons and everyone walks around on eggshells, wondering if you're going to change your mind at the enormity of it all, and whether the situation was really as bad as you painted it. Add children to the situation and you will also have comments made (both to your face and behind your back) as to the wisdom of your decision to leave. The situation is awkward and the longer it goes on, friends start to polarise to one side or the other. None of this actually helps either party, who in the main try to work out things as calmly as possible for the sake of the children.

I arranged a quick trip back to Glasgow to see my solicitor, met up with a wonderfully supportive friend and told her to start looking for a house, and began the process of separation. Thankfully, my husband made no real effort to save the marriage. Our marital home was sold quickly and within six months I'd bought a tired old-fashioned do-it-upper in a pleasant suburb of Paisley. Within a few months I'd landed a good part-time clerical job at a further education college, began a small garden maintenance business in my "spare" time and was again renovating. I persuaded my mother to baby-sit to allow me to continue with my degree with the Open University and burned the candle at both ends trying to be superwoman.

My mother insisted we visit twice a week when she would cook for us. Partly, this was to try and exert control over me, and partly to see the children and ensure they were eating correctly. So the children loved visiting, and I tolerated being lectured about my shortcomings. I have to explain that she was once a superb cook; making Middle Eastern food, curries, and could also bake really well, so food became more "interesting".

One evening a sponge cake arrived on the table sprinkled with sugar. This was odd. She saw my enquiring look and laughed.

'I forgot to put the sugar in the mix, so I just sprinkled it on the top. Eat it - it's fine!'

Kieran ate a big slice whilst Kate and I took small slices to be polite. It was certainly novel. I offered to bake next time, which was met with a torrent of anger, and I learnt just to accept Mum's increasing ire. She was now in her seventies, and my brother and I assumed she was turning into one of those old grumpy right-wing newspaper readers.

We reduced visits to Mum to once a week, as I was busy with assignments for my Open University course. However, she had other ideas, and I'd frequently return home after collecting the children from school or nursery to find her standing on the doorstep. She'd come by train and thought she'd "drop in". This meant she would stay till 10pm, then I had to pile the pyjama-clad kids into car seats and drive her home an hour away, as she wouldn't catch the train at night. She repeatedly asked for a house key, which I repeatedly refused. Eventually there was a "clearing of the air". I tried as calmly as possible to explain that, whilst I really did value her visits, they were spur of the moment, and with work and other commitments I couldn't always be there to meet her; that I had my own life and wasn't going to assimilate into hers. Her response was that she would take me to court to see her grandchildren! She stomped off to the train station and I let her go, furious at her response. My brother phoned a week later to complain of my treatment of her, but after I explained the other side of the situation, he complained that he too was getting it in the neck from her.

Back at my own house, my free time in the evenings was spent doing assignments, or stripping back countless layers of wallpaper and redecorating as the budget allowed. It took weeks to remove the old wallpaper, and then to repair and smooth down the surface. By the end of the process I was tired, but the periods of reflection saw me eventually regain my former self confidence. Six months later, I had stopped mourning my marriage, and celebrated completion of my penultimate Open University module. I was still a little raw and tender, but whole; and ready to move on in life.

Despite my mother's attempts to prevent me having any kind of social interaction (as I think she saw me as soiled goods and should be acceptant of my new status as a failed wife), I managed to secure an occasional mid-week night out to visit a local camera club, and have conversations with adults about a subject I enjoyed. The club had a mix of characters, some friendly and supportive, and some more than a little odd. A few of us made a habit of finishing the evening off at the local pub, where the atmosphere was more relaxed than in the club itself. I arrived late at the club one Wednesday, explaining that I had a slate missing from my roof, resulting in a small but persistent leak in my bedroom. Peter, who lived a few streets away from me, mentioned he had a tall ladder, and John volunteered to come round early the next evening, collect the ladder and sort the slate, which would save me spending a substantial amount of money getting someone in to do it. At the time I thought he was just trying to avoid the possibility of me falling spectacularly from the ladder, and leaving the children motherless, but he fixed the problem in less than half an hour and declared it no problem at all.

A short time later the delightful problem of a blocked toilet arose. The water level refused to drop, no matter how much I plunged the pan. As prime suspects, the children were questioned carefully and separately as to what the possible problem could be. Both denied they had put anything unusual down there. A professional firm was called and the plan was to jet blast the blockage away. I had to agree, and pay in advance, which I was unhappy about, but he assured me he was the expert and this would take a few minutes to sort. Less than 10 minutes later my whole bathroom was covered in sewage and the house stank. The blockage however, remained. I'm delighted to say that the "expert" performing this "simple" removal of the blockage was completely covered in the back-blast, and, after failing to clear the pipes (but refusing to reimburse me), had to climb back into his van covered in shit.

I had never liked the pink bathroom carpet and, after washing down the ceiling, window and walls, I rolled it up and carried it out to be binned. I left the window open all night and installed a temporary "Bucket of Shame" for the necessary functions. The next day, the smell had started to dissipate, whilst I searched the internet to find a solution. Determined to remain independent; I worked out that the blockage was further down the sewage pipe. In the inspection hatch at the end of the driveway the sewer was clear, so the blockage had to be between the toilet and the end of the driveway.

It took me a whole day to excavate the sewage pipe in the driveway with a pick and shovel. My aloof neighbours watched covertly from their kitchen window. Finally, I uncovered about three foot of pipe. I phoned John, who came straight from work with an angle grinder to cut open the pipe. He was furious at me for not phoning him before, and cut an inspection hole in the pipe. It was blocked. He fished out a piece of slate which must have come from the roof at some time, but the level of sewage dropped only slightly. He took off his shirt and reached in with his arm, groping around up to his armpit in sewage. Finally, he withdrew his arm, releasing the backflow of effluent, and opened his hand to reveal the blockage.

For those fans of the children's book and television series Thomas the Tank Engine, I can positively identify the blockage as Percy; the small green train from the series and a particular favourite toy of my son. John smiled and reminded me he was just a child.

Summoned and presented with the washed and disinfected evidence, Kieran tearfully owned up to being the culprit. He had been playing with it in the toilet, and swore never to do anything like this again. The pipe was repaired and the driveway filled back in.

John became a firm friend and regular visitor to the house, helping me to clean up the outside of the house and carry out DIY jobs I couldn't do myself. We took the children on photographic assignments and spent more and more time together; eventually acknowledging that our friendship was turning into a relationship. After a brief courtship and much silent rolling of the eyes by both our mothers we decided to take the plunge and get married. And that, as they say, was that.

With John's support, I then ran my gardening business full-time (with John working with me on larger jobs as I was so busy). I graduated from the Open University with a good honours degree in Humanities and finished renovating the house. As summer ended and autumn approached, I no longer had my studies to look forward to, the gardening jobs were slowing down and I had a lot more time on my hands. And so, inevitably, I started to have that dream again.

I blame the weather.

In Scotland, the weather dominates everything. Oh, Scotland is beautiful; with rugged, dramatic scenery, romantic castles and islands and a vibrant heritage. But the weather can be pants. It's frequently dull and wet and winter lasts for at least six months if not longer. You go to work in the dark and come home in the dark. It's depressing and monotonous.

Despite what the Scottish Tourist Board tell you summer is short, and although you get the odd week of brilliant sunshine and possible temperatures of up to 26 degrees you also get dull grey days and lots of rain, and on the West coast the inevitable midges. It's not all heather-clad mountains, Rob Roy and golf courses. It's also overcrowded towns and cities with overpriced property. It's a racking cough that starts in September and only disappears in May when the coughing has weakened your bladder muscles sufficiently to ensure that every sneeze, laugh or fart is accompanied by a squirt of urine. I was thoroughly sick of it all and far too young to start a meaningful relationship with incontinence pads.

On top of this we were outgrowing the house; and neither of us wanted to expand the gardening business and work even more hours. John's father had decided to retire and close his business, leaving John without a job and no fixed idea of what he wanted to pursue. He spent the autumn tidying up his own beautiful but neglected house in Dunoon on the Argyll peninsula and sold it immediately. Eldest Genius Child Kate was also reaching the age where we had to choose secondary education, and neither of the two available choices were going to be right for her. And at the back of my mind was the nagging voice reminding me that this was the time! This was the opportunity!

And so I examined rural properties, and the old dream came bursting back with vitality and colour and longing; challenging me to take that big step into the unknown.

'I have an estate agent coming round on Friday', I announced at dinner one night.

'Are we moving?' said Younger Child with wide eyes.

'Just for a valuation.' I tried to sound nonchalant.

Eldest Genius Child slammed her knife and fork down

on her plate and glared at me. I knew exactly what she was thinking; that we had only been living in this house for two years.

You could have heard a pin drop, but the train had left the station by then, was picking up speed along the track, and the sound of the engine (or it could have been my thumping heart) was all I could hear in my head.

2.
The Undiscovered Country

How many times do we say to ourselves, 'I wish I'd gone there/done that/told my father...?' We wish our life away instead of making even a small change in our journey and having the courage of our convictions. There is only *this* life. Time swiftly passes by and opportunity is lost.

I am, by nature, a secret planner; with all the thinking and planning done silently in my head. So, although things might appear to others to be spur of the moment or spontaneous, there has usually been a vast amount of thought behind them. I like to be in control where I can be and I am constantly reviewing plans and ideas as life's little changes occur and have to be accounted for. For major decisions I do imagine and plan for five years in the future. Every possible scenario; every possible obstacle. I picture and evaluate the possibilities on an invisible flow chart in my head. Therefore I can totally dismiss any suggestions that I 'haven't thought things through' and my position regarding taking risks is balanced and I'm not as impetuous as first appears.

Gene Roddenberry famously "borrowed" and reused the original Shakespeare lines from Hamlet (Act 3, Scene 1) to refer to the future as "*The undiscovered country*'" in the Star Trek movie of the same title. And life really is like a journey into the unknown; with twists and turns coming from nowhere and leading you goodness knows where. There is no dress rehearsal, so you need to pull your act together and get on with the here and now.

So, I did a bit of research on suitable properties on the west coast from Ayrshire right up to Oban, but soon realised that what I wanted was way out of our price range, or was very remote or located in the centre of a midge-filled hell in the summer months, and a snow-filled prison in winter. I had a new husband and children to consider, so moving had to fit in with so many wants, needs and expectations and planning had to be thorough. In the school holiday week in October we drove to Cornwall for a short break with the children, and I realised that this area ticked a lot of our boxes for relocation.

Although employment opportunities were far fewer and less lucrative, the weather was better, there was a smaller population, small schools, and a new opportunity to live more rurally. We quickly realised that the properties we liked in Cornwall were; like those in Scotland, out of our price range, unless they were really run down, located in old mining communities or in the very far West of Cornwall at the very edge of the known world. And then there were properties with an Agricultural Occupancy Condition on them. They stood out like tiny grains of gold in river mud. I absently searched the internet for basic information about them and, on our return home, had a telephone conversation with my long term solicitor. I had known Jim for years, and we understood each other and I could certainly tell him anything, and rely on his opinion and advice.

'So, tell me exactly what an Agricultural Occupancy Condition is', I asked.

'Why do you want to know? You're not a farmer. You can't fulfil this. Are you moving house *again*?'

I could hear him throw himself back in his chair.

'I'm just looking at options.'

‘What *kind* of options? You’re going to move again, aren’t you? Where to this time?’

‘Possibly Cornwall. The weather’s better. Smaller schools. Less drugs and bad influences for the kids.’

‘Drugs are everywhere,” he stated with a snort, and tried to change the subject. ‘How’s the new husband?’

‘Much better than the old one, thank you. So what is an Agricultural Occupancy? Why can’t I fulfil it?’

‘It’s very complicated; but basically, an ‘Ag Tie’, as they are usually called, is a condition of use put on a rural property by the local planning department, that it must be occupied by a farmer, forester or someone working in horticulture. It’s to stop townies running the countryside. *You* cannot fulfil it, even though you have a gardening business. Gardening is *not* horticulture. I’ll email you some stuff…- if you are seriously looking.’

There was a slight pregnant pause as Jim waited to see if I took the bait.

‘Yes…- I think I am seriously looking,’ I decided.

‘So, meantime, do you want me to handle your house sale? I’m assuming, knowing you; that you’ve had a valuation done?’

This is why Jim and I get on so well. He’s no fool, and I don’t believe he thinks I am either. Although he might not agree with my house buying lifestyle, I’d never made a property mistake and we had a good working relationship. My current house was the seventh I’d owned and my next move might be my last; I hoped rather than seriously believed this might be the case.

We visited Cornwall twice in spring and confirmed in our minds that Wadebridge was our chosen area of choice. There was a choice of primary schools, a good secondary school, and everything we thought we needed. There was even a small airport at Newquay that could take us to Glasgow if needed. We visited a quaint local pub in one village for lunch as a break between house viewings; having now decided to see what we could get for our money in this area.

'Bleddy emmets keepin' locals out o' t' villages. Bleddy 'oliday 'omes!' roared an old man, catching sight of my hand full of schedules. And then seeing Younger Child looking wide eyed at the outburst, the man smiled at him, 'Alright, my 'ansom?'

Oh great, I thought, *outsiders really do get a warm welcome in Cornwall.*

'What language was he speaking?' whispered Eldest Genius Child squeezing round a table to sit and peer at the man from over a menu.

'Cornish!' said Younger Child in an attempted Cornish accent.

'Enough! Choose something,' I said thrusting the menu at him.

'What's an *emmet*?' asked Eldest Genius Child.

'Mum', said Kieran, 'the toilets here are weird. In the gents you actually pee on a wall.'

'It's a trough,' I corrected, glad to change the subject.

'No, it's not,' said John. 'Probably the same toilets that have been here ever since it became a pub, I think. Rustic isn't the word.'

The ladies loo, I discovered after lunch, was just as rustic, with a hole in the roof and an umbrella propped in the corner to shelter from the rain when accomplishing the necessary. Oh the joys of rural life.

When we returned home, we cleaned and spruced up the house and put it on the market in June, and almost immediately received an offer at full asking price. I was standing in mud-covered jeans with wellies half full of rain, laying a new lawn for a client in the south-side of Glasgow, when the phone call came. It was a typical Scottish summer day, and John and I had just finished laying turf and were wet, filthy and tired. Could we complete by the end of October?

Oh, if my mother could see me now, I thought, before taking a deep breath and saying, 'Yes.'

*****

Time was against us and we squeezed in as many viewings as we could into a weekend. John flew from Glasgow to Newquay, hired a car and drove around to do the house-buying equivalent of speed dating. He could tell in a few minutes if the house was a "possible" or "reject". Wrong location, badly renovated, too small or too expensive were the main reasons why there were no second viewings. Finally, the only possible properties left were those with Agricultural Occupancy Restriction's. I had extensively read up on them, and understood that we would have to either have agricultural jobs (not easy in a small, close-knit community that were unwelcoming to outsiders) or start our own business to fulfil the tie. Having already been successfully self-employed, this held no issues for me. Looking at our lack of knowledge or experience in agricultural skills, a plant nursery/smallholding seemed the best business to go for. We were to travel down to see some of these AOC houses in our next flying visit.

It really was a rush, trying to see everything in a weekend and give ourselves enough time to drive the kids the long nine-hour journey back, for school on the Monday. We quickly dumped our things at the holiday cottage and sped off to try and locate the first property. It took us ages to locate as all the lanes looked similar, and the local Young Farmers appeared to have switched the old fashioned white finger-post signs and we went round and round in circles for at least forty minutes before we realised. The narrow lanes were hemmed in with tall Cornish hedges obscuring any chance of trying to see a landmark, such as a church spire, and we spent ages reversing to passing places and braking suddenly when meeting other vehicles. Instead of enjoying the scenery and relaxing, it was like driving in World War One trenches, and made us both irritable.

The first house was located on a windswept barren coastal moor, with schools and neighbours miles away. Raggedy thorn hedges blossomed with tattered plastic carrier bags and an ever-present wind. The house was grey and unfriendly with its backside against the wind and hunched low, clinging to the ground. The agent told us the owners had tried to run a palm tree nursery but it had failed. The fields were full of rushes and pools of standing water, with no livestock to be seen, and I wondered how on earth farmers could ever make a living here. John and I couldn't drive away fast enough, much to the children's delight.

The second house was in the centre of a high moorland village that was regularly cut off in the winter. The land was stony and thin and again the site was windy. The trees cowered behind dry-stone walls and were stunted like wizened dwarves. This appeared to be a predominant theme in properties in North Cornwall. The access roads to the village were narrow and steep and the stone banks edging them were scarred and damaged with scatterings of orange indicator glass or pieces of number plates. John decided, after a visit to the village shop for chocolate bars that this was a firm 'No'. He said it reminded him of '*those weird villages you see in old vampire or werewolf movies*'. This resulted in not just the particular property being scored off the list, but the village too.

Then we arrived at property number three. It was a 1980's bungalow, a mile outside Wadebridge. We drove up a pitted farm track, past fields of Miscanthus, a tall bamboo-type crop originally subsidised by an optimistic European Union for growing as a bio-fuel, but sadly lacking any real market in Cornwall, and being eventually used as animal bedding. It swayed in the breeze, waving little elephant-tail plumes and welcoming us with a sound like the faraway sea. Further along the lane was an apparently redundant farm, which had largely been turned into small units and lockups. The bungalow was tired and sat in an acre of grass. It wasn't the most attractive house and certainly would never be a "chocolate box dream". I kept reminding myself that anything in a house could be changed except its location. And the location was good.

It was immediately obvious that despite being an agriculturally-tied property, it had never been a smallholding. It was too large for a typical farm workers' dwelling, especially with the huge but neglected swimming pool and we grew increasingly sceptical that it had ever been a true "home". Inside felt bare and cheerless. There was no atmosphere, no echo from past owners. It appeared that the house had been built, kitted out very basically and then left. However, the rooms were large and the living room had an enormous picture window which looked over the fields and down to the Camel Estuary. The property schedule highlighted the "estuary views", which would be pleasant; when the tide was in. It had what estate agents optimistically called "possibilities". At that moment I could certainly see the possibility of spending a large amount of money turning this into my dream property.

The estate agent slipped the fact that the owners had been renting the house as a holiday let for some years and, having failed in a previous attempt to lift the Occupancy Condition had been forced to put the house on the market some months previously. Having done my homework on agricultural ties before coming down, I imagined the owners were trying to market the property for a year or so, hoping no-one would put in an offer, and they could then try to legally remove the tie and sell up at the full market value.

'Ideal for updating,' said the estate agent cheerfully.

'Money pit is more realistic,' announced John.

'Complete new central heating system needed for starters,' he told me, peering inside the burnt out Aga in the kitchen. John and I exchanged more than a few looks with raised eyebrows and a realisation that this would be the biggest restoration that either of us had taken on before. I was mentally adding up huge sums of money in my head as I walked around inside.

'No pigs at the farm next door anymore. Foot and Mouth closed the business,' said the agent. So the neighbouring farm had just decided to stop farming altogether? I imagined that the fields surrounding the farm, and this house, were now rented out to other farmers. We looked round the land and house again quickly, whilst the kids bickered in the back seat of the car.

'Lots of viewings,' the agent stated, confidently at the end of the tour.

I imagined this to be far from true, but admired his half-hearted attempt to make us think the house would soon be inundated with offers.

'But none who could fulfil the occupancy conditions?' I countered with a knowing smile.

'So what sort of farming do you do then?' he asked conversationally, but obviously trying to ascertain how I was going to fulfil the requirements of the Agricultural Occupancy Condition. After all, he wanted to sell the place; not just waste his time showing round people who couldn't possibly fulfil the terms of the restriction.

'Market gardening and plant nursery', I answered as confidently. 'I'm thinking of moving my business down here.' I sounded very smooth, I thought.

Eyebrows shot up as he sniffed the possibility of a sale.

'Oh, *right* then. Well, because of the restriction, it's very reasonably priced for a large property in a desirable location.' Suddenly, the charm was turned on as he realised he was finally showing someone round the property that could fulfil the restriction. He wasn't keen on the idea of marketing a property for a year and getting nothing in return – he was thinking of his commission. This could be something in our favour. I wondered how long it had been up for sale and if there was a deal to be done.

'Hardly a bargain,' I said tartly. 'It's right next to a disused farm with dilapidated buildings, let out to a number of "interesting" small businesses, and the house needs completely renovating. The price would have to be very negotiable, despite the tie.'

I waited for a moment to stress my next comment.

'We've agreed a sale on our house and are ready to buy something as soon as possible - *if* the price is right.'

'Well, naturally, I'm sure I can help in the negotiating process.'

Yes, I'm sure you can, I thought.

We had played the game of estate-agent-bluff well; letting him know we were interested, but not at the price the vendors were asking. There was also the bait of a further two acres adjoining the property, if we wanted it. At the time we were both of the opinion that one acre was quite sufficient. It seemed a huge space to us as inexperienced would-be smallholders.

'We'll go and have a chat and get back to you,' I said, heading to the car.

We drove down the lane, taking in that spectacular view of the estuary again, and headed off to find a pub and discuss the possibilities.

'Are we going to buy that house?' from Eldest Genius Child.

‘I don’t know. Do you like it?’

‘It’s alright, I suppose, but we’ll be country bumpkins,’ she said gruffly, reminding me of my mother.

‘Oh Aarr!’ said Younger Child, mimicking the local accent and laughing in his sister’s face. She pushed him away.

‘Do I have to share a bedroom with *him*?’

‘No, you can have the biggest bedroom, *if* we buy it.’

‘I’ll miss all my friends,’ she sulked.

‘You’ll make new ones. Nicer ones,’ I reflected. This much, would actually be true.

‘Can I have a pony?’ Kate asked.

‘Can I have a tractor?’ Kieran asked.

‘No.’

‘...Maybe a tractor,’ John added as an afterthought.

*****

A lot of negotiation, time and sadly, more than a hint of hostility from the existing owners went past. They refused to move much on the price, which meant we had to take out a small mortgage to secure it. Then there was an inter-family dispute from the seller’s side, but finally a price was agreed, the date was set and contracts signed. We would collect the keys in late October and made plans to coordinate the huge relocation. Very soon we would own a three bedroomed bungalow with just over an acre of land almost a stone’s throw from the beautiful Cornish coast.

It wasn’t only a huge logistical nightmare; after all Cornwall was about nine hours’ drive in a fast car on a straight run with no traffic. It was moving the complete lives of four people. It was leaving family and friends, schools and jobs, the place you were raised, and moving almost to a different culture, to start a completely different life. I was ready for a change, despite concerns about the children making this huge move, and trusted in John’s quietness that this was a sign he too was looking at it as an opportunity.

In the back of my head I wondered even then, if we were really capable of pulling it off, of making a success of a completely new way of life with the few agricultural skills we had, but firmly shoved this thought to the back of my mind. I don't know who was more horrified when we returned home and told them; my mother or my solicitor, but we had made the decision and that was the end of it.

And so we packed all the furniture, books, tools, toys and everything we thought we would need and filled a removal van. John went round to his dad's shed and packed up all his dad's blacksmithing tools and an anvil which were no longer any use to his dad, and filled our gardening van literally to the top, squeezing something into every last space. For me, this was just moving house again; another notch on my home-buying record. For the children it was more emotional. They were leaving their home and friends, journeying into an unknown future. John had to leave his parents and sister, which was very hard for him and an indication of his commitment to me and my dream. I hoped it would soon become his dream too.

On our last evening in Scotland my mother made us a farewell meal of creamy chicken and mushrooms with rice. This was the children's favourite and she was subtly reminding us what we would be losing. As we cut into the chicken, it was raw and I quickly told the children not to touch it. I didn't really want it to be a farewell meal in that sense of the word! She; of course went ballistic, and blamed me. It was my fault. I'd upset her so much she had forgotten to defrost the chicken. I rolled my eyes. *Here we go again*, I thought. This had all happened because we were taking her grandchildren so far away, she said. I'd never considered her at all in any of this, she said. It was just as well my father wasn't alive to see this, she said. We left to prevent an argument and drove home, stopping en route to get fish and chips and I rejoiced in the fact that I would soon be free of my mother's dominating and increasingly selfish behaviour.

The next day the convoy set off with John and Kieran in the removal van; Peter and another friend in the gardening van, and Kate, me and the dog with an assortment of plants and boxes of important paperwork in the car. I walked round the empty house for one last time, feeling a tinge of sadness, but more a tingling sense of the adventure ahead, finally locked the door and the past behind me, and looked towards the future.

Kate and I stopped overnight near Bristol as I just couldn't drive any further, and when we arrived the next day, John and his friends had already unloaded the van. The children excitedly helped sort their things in their new rooms and went exploring. We all fell into beds exhausted that night. The next day we did a proper tour, and stared a lot at the enormous, elderly swimming pool which was dark green and foreboding.

'Good to have, in case the house catches fire,' John said cheerily.

'Well, we certainly won't be using it to swim in,' I said, 'and I bet it'll cost a lot to fill in.'

'It's big enough to be a building plot,' he mused thoughtfully, '*if* we had the money.'

'Maybe we will, in the future.'

He turned and looked at me with a smile.

'Most women set their sights on a modest goal in life. I don't think you *are* like most women, and wouldn't be surprised at *any* goals you set.'

Oh, how I wished I felt as confident as he obviously thought I did.

The land was almost flat, contained within a Cornish wall with lots of overgrown sloe hedge alongside the access lane.

It was perched on the top of a small hill, with the remains of an iron-age hill fort just across the lane in the field. From my research, we knew this used to be called Kelliwig rounds. The lane from the main road to our house ran straight through the centre and out towards the village on the hill nearby. Some internet sites thought this was the fabled site of Camelot, but more realistic archaeological investigation revealed it had never been inhabited and was more likely to be a traditional meeting place, although it was fortified by two rings of raised mounds, possibly surmounted by a wooden palisade. Through disuse and neglect the rings were only visible in winter now, overgrown with sloes and bracken. To the front of the house was a lawn, with a usable half acre of land to the rear, and a few trees along the perimeter as shelter to the South West. This was to be our working area, where we would site a polytunnel, plant a small orchard and perhaps keep some livestock. Even then, I started making a mental shopping list.

That night, after taking our friends for a mediocre meal at the run-down local pub, John and I left the kids inside watching television whilst we walked to the gate for a little quiet time on our own. In the packing up, driving and unpacking we had hardly had two minutes alone together. We leaned together on the gate and looked out at new skies.

Like most of North Cornwall, there is hardly any light pollution once you are away from the towns. It was a beautifully clear night with no moon and the stars were flung across the sky like sugar on a dark blue tablecloth. There wasn't a breath of wind, and somewhere quite close we heard the scream of a barn owl. It was the only sound in the deafening silence. The Milky Way stretched from the South behind us to the North and the sea.

We had made the incredibly huge decision to move our home and our lives right across the country and were embarking on a whole new way of life. Was I nervous? Hell, yes! We'd left a good business, taken the kids from a school where they were happy, and left all family and friends behind. In a way, I felt like an explorer; thinking this must have been how the Highlanders felt after the clearances, when some embarked on ships to make their new lives in the USA and Canada. Although I was optimistic, I wondered what our chances of making a success of it all would be.

'It's a crap house,' John finally said, 'but you'd kill for a location like this.'

'It's a project. We'll make it better. It needs love.'

'It needs a millionaire.'

The next day, John left early to return the large removal van, and his friends to Scotland. The kids and I sorted the furniture as best we could and used the draughty wood burner in the living room for the heating, as the electric storage heaters, we discovered, were disconnected. We had no wood or coal, and spent the day searching the area to buy logs and kindling, ending up on the beach where we collected some firewood washed up on the tide.

Nothing will make you more popular with your children than dragging them away from their friends and moving them to the other end of the country to a dilapidated house with no means of heating. Tempers were frayed and the children sullen. School was another hurdle for them, trying to integrate into very small classes of established friendship groups, but they did their best. I was examined closely at the school gate at collection time and was obviously the "new person" to be discussed and gossiped about.

Fortunately, John returned shortly with an LPG boiler, radiators and piping and began installing the central heating we badly needed. All we were waiting for was the LPG tank and connection, which we were informed would happen in January. We began to understand the Cornish phrase ‘dreckly’, which could mean anything from a couple of days to months, which to us was annoying but had to be accepted. But it was *fine*. We were Scottish and used to cold weather, and it was much warmer in Cornwall than Scotland.

On November 24th it snowed. I mean it *really* snowed. Despite being Scottish and used to occasional falls of snow, this took us by surprise. We weren’t the only ones. Cornwall ground to a halt, with the county totally unprepared. There was no grit for the roads as Cornwall Council hadn’t bought enough (because it *never* snows in Cornwall). Local drivers were inexperienced and there were abandoned cars and crashes everywhere. At lunchtime the snow was six inches deep and the primary school shut; forcing me to walk the two miles to the village to collect the children. Luckily, on the return walk we were offered a tractor ride back to the relative warmth of the house. The kids were thrilled and I was grateful for the generosity of the farming community.

By tea-time the A30; the main arterial route through the county joining Cornwall with civilisation, was shut with hundreds of cars, people and a coach load of school kids stuck out on the moor in treacherous and freezing conditions. We fed the wood burner continuously with logs and tried to encourage warm air to circulate through to the rest of the cold and draughty house. Television was watched whilst we huddled with ski hats under duvets. In case I wasn’t aware of the fact, my mother phoned to tell me it was so bad that Cornwall and the weather were on the six o’clock national news. She barely managed to hide the glee in her voice as she told me she thought I’d made a terrible decision.

3.
The Circle of Life

John spent weeks crawling around in the void under the floor, installing pipes and connecting radiators to the new boiler. He emerged every evening filthy and tired, but never ever said one word of reproach. I was amazed at the existing skills he had and how quickly he could master new ones as required. Kieran offered to help to go under the floor with him and pass him items, but when John told him the size of the spiders under the floor he quickly changed his mind. Whilst we waited for the LPG tank to arrive and be connected, we removed the old loft insulation, replaced it with new, and doubled the depth of it. It made an almost immediate difference to the heat in the house. We stuffed the old insulation around the bath then replaced the panel, realising this would keep the water in the bath hotter for longer. A few sheets of insulation boards were cut to size and John took them under the floor to jam between the wooden floorboards under the bathroom to further insulate it. Cavity wall insulation soon followed when we found a local grant scheme offering to do it free as part of a government initiative.

Finally, in January, we had proper central heating. For the first time ever, the bedrooms at the end of the bungalow were warm and cosy. At last we could look at the house and land and decide what we needed to do. The plan; right from the start, was to live as sustainably and cheaply as we could. At this time neither of us had heard of permaculture, but we knew we wanted to have a smallholding where we worked with nature as much as possible, aware that our limited land space was going to demand a symbiotic system. The house was finally coming together; the central heating was complemented by a second wood burner to replace the defunct Aga in the kitchen, which had taken a week to break up and remove.

That winter was hard work trying to make a home, integrating kids into a new school and village and making friends in a small community. The house was isolated which didn't help. Living rurally, surrounded by fields meant no neighbours with children, no park to run to, no football in the street to help build friendships. The village was a good two miles away up a very steep hill, so bicycles were out at the moment.

Luckily the village had a park and, as spring approached, we made the most of meeting new friends after school, which helped the children settle. During the day, I tested the soil and discovered our high water table and the differences in the soil – shallow and full of stones and shards of flint or slate in the front garden and deep and loamy in the back paddock. Our plans to put the orchard in the front would now have to change as the soil was too shallow to plant fruit trees. A shelter belt would also have to be planted at the rear of our paddock. The prevailing wind was coming from the South West and any fruit trees and our planned polytunnel would need a bit of shelter as we were on the top of a hill. I managed to source a supposed miracle crop called Mammoth willow, and planted a double row of staggered sets in the far South West corner of the paddock. A previous attempt at planting some ash and hornbeam was shelved as the planting holes filled with water when I reached a depth of nine inches. So, water-loving willow was the only real option. As they grew the idea was to coppice, to produce kindling and to thicken the clumps of willow, to eventually form a thick green hedge two metres tall.

John had set up a workshop in the large agricultural shed and reluctantly found a temporary job just outside Wadebridge, to stem the flow of capital from our dwindling bank account. However, one of us had to try and fulfil the terms of the Agricultural Occupancy Condition.

This usually means being a farmer, farm worker or a retired farmer. Buying a handful of sheep or horses just won't fulfil the terms of the restriction and you can expect a visit from the Council's Enforcement team to check you are indeed complying with the condition. Our solicitor in Wadebridge had taken pains to make the situation very clear. I had decided to start a plant nursery as we had to show that we were fulfilling the wording of the occupancy restriction, which was: *The occupation of the dwelling shall be limited to a person solely or mainly employed, or last employed, in the locality, in agriculture or in forestry, or a widow or widower of such a person, and to any resident dependents.*

Thankfully horticulture and market gardening were deemed to fulfil this. So, the dwindling pot of cash from the sale of our Scottish properties was invested in equipment and supplies to start a small plant nursery, specialising in hardy ornamental perennials and shrubs. This seemed the safest option. I had years of experience growing plants, and Wadebridge was a small tourist town a few miles from the coast and other tourist villages. Hotels, holiday houses and the like would soon be buying plants from us. What could possibly go wrong?

We also bought our first livestock. John made a hen house from timber and weld-mesh, and we bought four Light Sussex pullets or young hens. Of course, being completely new to all this we gave them names. Shortly afterwards, we also acquired a "free to good home" Light Sussex cockerel, called Dave. We could eat the eggs, hatch young birds and any cockerels could be fattened and eaten. Dave was really friendly, his owner said. And he was - until he was about six months old and his hormones kicked in. Then Dave turned into a mean, vicious Berserker.

From his pen he would watch you approaching, head to one side, waiting for his moment, and then rush at your legs, pecking and scratching. As his attacks became more vicious and sustained, and his spurs grew longer, we began to carry a stick when venturing into the pen. Crunch time came when Dave's spur burst through the side of the feed bucket, which could have as easily been any one of our legs! I'd heard enough horror stories about cockerels by this time; especially from one of the old boys in the village. Rex recounted the story of a Light Sussex cockerel which had flown into the face of a child and blinded him in one eye. Tall tale or not, enough was enough. It was time for Dave to meet his Maker. Everyone I asked graciously refused to do the deed as his reputation was ferocious. But his days were numbered, and I fed him copious amounts of maize and pasta to fatten him up for D-Day.

I watched internet videos of chicken dispatch until I felt ready to do the deed. I then phoned the husband of a friend and asked him to come and show me how to prepare a dead bird for the oven. There was no way I was killing the bird and not eating it! So, after trapping Dave in a corner and smothering him with a large jacket to protect myself, I took him, shrieking his angry head off, to the side of the shed. I kept tight hold of his legs and held him face down on the ground, put a thin bar of steel over his neck and then placing my feet on the bar either side of his head, I sharply pulled his legs up and forward, like the video had shown me.

I felt the neck go. He was dead. Elvis had left the building.

But his eyes were still blinking and then…the wings started! My God, the *ferocity* in the beating of those wings! If I hadn't kept hold of his legs I think he would have taken off into the air. After about half a minute (though it seemed so much longer) it finally stopped and I lay him on the ground. I was shaking, sweaty and my heart was racing like an express train. I headed inside for a glass of wine, to steady my nerves, before Brian arrived to show me what to do next.

'You managed it then?' he asked when he arrived.

'Yeah, piece of cake really,' I lied.

'Proper job! You gorra bucket then?' he asked, picking up the bird and carrying him to the table in the workshop. I indicated the waiting bucket and watched as he made a slit in the vent area, enlarged it, put his hand in and pulled out everything inside the bird, and dropped it in the bucket.

'Just pop your hand in so as you know what it feels like when he's empty,' Brian said with a badly concealed smirk.

Gingerly, I slid my hand in slowly, and as I did so, a loud noise like a sigh or a groan emanated from the bird. I yanked my hand free with an exclamation.

Brian was snorting with laughter at my inexperience.

'No, he's definitely dead,' he laughed, 'Iss just trapped air.' He wiped the knife and his hands on a towel I handed him.

'I'll be off then,' he said heading to his car.

'But what do I do *now*? How do I pluck it?'

'Just pull 'iss feathers off. Against the grain.'

And then he was gone. Leaving me to deal with the corpse. Didn't tell me what to do with the head, the feet, anything. I looked in the bucket. All Dave's insides were in there in various hues of purple, red and beige. I pulled up an old chair and started to pluck, trying to get the feathers in the bucket. It was a slow job. Then it was off to the wood shed with the axe and off came Dave's head and legs. The carcass before me had been very much alive and kicking just an hour or so before. Now, there was little to identify this naked body before me with the vibrant and violent creature that had been our adversary.

Life is fleeting. But this was the life we had chosen, and Dave's death would not be in vain. It would, with the addition of vegetables, feed all four of us and he had had a good life all things considered. I wondered how we would feel about taking larger animals to be slaughtered, as we planned shortly to buy sheep. With this thought in my head, I took the carcass inside and gave it a wash in the sink, and then placed it on a roasting tray, surrounded with some peeled carrots & potatoes. It vaguely resembled a supermarket bird, but not as plump. Then, I popped the tray into the oven, and left to collect the kids from school.

Dinner that night was memorable. Out came the bird, which I'd had to baste a lot, and the normal chatter at the table died abruptly.

'Is that Dave?'" said Kieran with eyes as large as saucers.

'I'm not eating that,' Kate pronounced, crossing her arms.

Determined to set an example, John and I took a leg and thigh each. The meat was quite dark, almost gamey. And chewy. The legs were very chewy. I cut off the breast meat for the kids and popped it on their plates.

Kieran dived in first.

'It's very... *chickeny,*' he said, 'the nicest chicken I've ever had. It tastes just like *real* chicken.'

We replaced Dave with another more docile cockerel, purchased our first 24 egg incubator, and John built a small indoor hatchery in a small shed. As we were hatching lots of chicks with a ratio of 50 males to 50 females, we had an ongoing surplus of males to fatten up, and a naturally produced ready source of chicken. We had also bought some Buff, Blue and Lavender Orpingtons, a couple of Legbars and a pair of Marans. Soon we had a lot of eggs of a variety of colours; white, cream, light brown, chestnut and blue and green (from the Legbars) and I had to decide quickly how to prevent us having an egg mountain in the kitchen.

I decided that to sell boxes of coloured eggs would be a way to pay for the rising amount of chicken feed we were using, and use up the surplus of eggs we were producing. The official paperwork was quite easy, the local Egg Marketing Inspector helping with everything. When I told him I intended only to have about thirty hens, he coolly informed me that in the half acre paddock, I could comfortably have a couple of hundred, and that this was the only way I'd make any money from selling eggs! The local shop-cum-post office in the village was soon stocking the eggs and selling them as fast as they arrived. In fact, on "egg day" they were waiting for me at the door!

I'd bought some Buff Orpington pullets to add a little colour to our monochrome hen flock, and then found a lovely Buff Orpington cockerel to go with them. These gentle, but daft birds would breed and provide us with new pullets to replace the old ones and young cockerels to fatten for the table. The few chicks we allowed the hens themselves to hatch that summer had a bit of a sad end. One was squashed by mum lying on top of her, one drowned in the deep water container we had placed in the pen, and two were taken by magpies. John refused to let me leave the hens with fertile eggs as a result and we used the incubator only for future hatching.

That summer we had the worst spring and summer weather in Cornwall for years. It was cold, and rained and rained for weeks on end. Just like being back in Scotland! We both lost our tempers on occasion, as we had been working very long days and the bank balance was dwindling quickly. Many young plants failed, and customers were few and far between. I did a few farmers' markets, selling trays of plants, but again, holiday makers didn't take things home with them, and locals weren't buying. It soon became apparent that Cornwall is a county of two types of people: the rich and the struggling. The rich included retirees who had flocked to Cornwall to spend their twilight years, and holiday makers who wanted a British holiday but with incomparable scenery, sunshine and good restaurants. Then there were the locals who, in the main, had to contend with high house prices and very low wages if they were not on benefits. Garden centres made their money from the tearooms and gift shops, whilst people bought plants from DIY stores and cut price outfits. Our expected sales to holiday-home owners and hotels never happened. We couldn't provide the range of plants or the cheap prices required to compete, but we struggled on. It was a bitter lesson we had to learn and quickly.

So, we explored how to best use the land and the facilities. We had between a third and a half acre at the rear of the house, which was the domain of the chickens, so we fenced it off to make a neat little paddock and discussed having a few sheep with the smallholder group we had joined. The children and I went off to see a breeder and she showed us them up close. Yes; I've seen sheep before, in fields, on the telly and at agricultural shows, but when you're invited into a pen to actually handle one for the first time it's a different matter. Suddenly, they were huge. Almost the size of a Shetland pony! The children stood on the other side of the pen and watched me with amusement, performing a game of catch with one particularly nimble and strong ewe. I like to think they were impressed, but Kate's face gave me the strong impression she thought I was being totally ridiculous.

Very soon we had purchased and installed three yearling Lleyn ewes, and arranged for them to go off in November for a romantic couple of weeks to a ram belonging to a friend in Devon. We had chosen Lleyn's as a breed as they were hardy, good mothers, and had a tendency to lamb easily. We had also decided to make an offer for the extra two adjoining acres that we had been told about when we bought the house, which would house our soon-to-be little sheep flock. This way I could think about using the paddock purely for an orchard and avoid sheep damage to the trees.

I hired a scratter, a rack and cloth press for the weekend and we managed to press a small amount of apples from our established apple tree and pressed around 100 litres of juice which we intended to try and make our first batch of cider. It was great fun, if horrendously messy, the shed reeked of apples, and we stored the plastic drum in a cool corner, checking on the quietly popping and fizzing contents once a week. By March it finally quietened down and we delightedly decanted the clear golden liquid into recycled beer bottles. Beginners luck had blessed us and the strong, dry cider was delicious. I was hooked and decided we needed more apple trees to add to the complexity of the flavour. When we had some money, this little experiment needed to be expanded.

One spring morning after dropping the kids off at school, I stood watching the sheep turning all that lovely grass into meat, and noticed that one didn't seem particularly bright. I walked slowly towards her, waiting for her to trot away, but she just stared at me. There was a small dark patch on her fleece just above her hip and, as she allowed me to touch her I manoeuvred her to the fence to hold her and take a look. Parting the fleece showed me a couple of maggots. I groaned, as I recognised flystrike, and left her to drive down to Wadebridge to our agricultural suppliers, where I bought a flystrike treatment. I phoned our friendly local sheep farmer friends who suggested I take her indoors and have a good look at her, cutting away any fleece to see the extent of the attack. Back home, I half pushed, half pulled her into the shed, where the lambing pen was in place, and proceeded to clip away and discover more and more areas where the dreaded flies had been. After applying the treatment, soon maggots were emerging from the skin in their hundreds and falling to the floor. It really was like a scene from a horror movie and I felt sick. So many maggots, and I had only noticed a small area of discolour on the fleece. I clipped more wool away and soon the poor sheep was fleece-free, looking like Sweeney Todd had clipped her. The maggots were everywhere, popping out of the skin all over the poor animal. The wretched sheep was so weak and uninterested she let me clip all the wool off her, using only a pair of sharp scissors. I applied more treatment all over her back, legs, shoulders and neck. This was the worst bit. Under the fleece the skin was almost raw on her poor neck. She looked awful.

With tears pouring from my eyes, I kept apologising to her as I found a broom and swept up the millions of crawling, seething maggots. I found an old feed sack and scooped them up with a small dustpan into the sack, wretching at the dirty, evil creatures. The sheep just stood and the maggots kept falling from her. I made up the lambing pen with clean straw and put a bucket of water and a bucket of pellets sweetened with some molasses to try and tempt her. By this time, I'd also spoken to the vet who said I really would have to see if she made it through the night as it sounded like a horrendous attack. She didn't sound hopeful. When John came home at the end of the day his face was grim as he looked at her.

'Was she alright yesterday?'

'Yes', I said, 'that's the weird thing - how could there be so many maggots in such a short space of time? Or was she already infected and just not weak?'

'She looks awful. Well, there's nothing more we can do till tomorrow and see if she makes it.'

We left her for a few minutes whilst we went to catch

and check the other two sheep for any signs, but thankfully they were clear, and we dosed them with the preventative treatment there and then.

The maggots appeared to have all come out. No more emerged and John took the feed sack full of the horrid things round the back of the shed and set fire to it, burning them all. We turned the shed light off and went to bed exhausted. I think; if I'm honest, both of us felt she wouldn't be alive in the morning, and I blamed myself for not being vigilant. I lay snivelling most of the night at the memory of the raw patches of skin, with my hands blistered and aching from clipping the sheep with a pair of kitchen scissors.

The next morning we opened the shed door with trepidation to find her looking as awful as the day before, but standing. The food and water had gone, so at least she had eaten something. She looked very weak and a good look at her skin showed some of it was healing. I replaced the food and continued with the day's chores. She improved daily and at the end of the week was back out in the paddock. A hard lesson had been well learned.

A week later I had a surprise visit from the Council Enforcement Officer, which was the icing on the cake. It was pouring with rain again, and I was already outside and pretty wet, having just fed the poultry, and was collecting eggs.

He didn't even get out of his car but pumped the horn. I walked over, and he wound the window down just enough to talk through.

'Yes?' I asked, as only a woman with water running down the inside of her jeans can.

'Mrs Turnbull? I'm the Enforcement Officer. It's my job to ascertain how you are fulfilling the terms of the Occupancy Restriction. I can see from the sign that you are a plant nursery. Can you tell me if you also have a part-time job? Say, in a supermarket or elsewhere?'

I was absolutely stunned, and indignant.

'Absolutely not! I spend all my time here. We're a plant nursery and have livestock!'

'And can you tell me how you make your living then? Surely you don't expect me to believe that your... "enterprise" here supports you?' he said with an ill concealed sneer, writing down my replies on a clipboard.

Furious, it took enormous control to swallow down the urge to swear.

'We moved here less than six months ago, as I'm sure you know, and are working very hard to make this business work. It's early days. Now, if you don't mind, I'm soaked, when I could be getting back to my work.' I stomped to the shed, and waited till he drove off.

Angry and very upset, terrified of losing hold of my dream so soon, I comforted myself with the thought that if John had been here he would have pulled the horrid little man out through his half-opened window and punched him for making me stand in the rain, answering damn fool questions. It highlighted the issue that we would have to be careful what we did. We assumed; correctly, that he could turn up without warning, at any time again.

The problem is that fulfilling an Agricultural Occupancy Condition (AOC) is difficult, as few people can comply. This makes the properties harder to sell, and consequently they are worth about a third less than a similar comparable property without one. After all, many people want to live in the country, but realise that prices are high. A house with an AOC is cheaper, but harder to comply with, and if the Council suspect residents are not complying, they can issue an Enforcement Notice forcing you to prove irrefutably that you do indeed comply. If you can't, they can force you to leave. This is particularly hard for those people who don't want to be employed in agriculture, but want to live a simpler, alternative lifestyle. The system does not accommodate such people, and the planning system actively discourages and penalises those who dare to try. In the years to come, we met and became friends with literally hundreds of people who wanted to have what we had or similar.

We found out how much the ex-owners resented us fairly soon. A water bill arrived for around eight times the amount we usually had to pay. We quickly visited the meter which was located a quarter of a mile away through the field in front of the house only to watch the numbers on the meter spinning round. We obviously had a leak somewhere. John quickly turned off the supply while I phoned the water company to find out what could be done. The next day, their expert arrived with listening poles and divining rods. He could hear the leak through the listening pole when it touched the pipe at the stop cock, but couldn't tell us where the leak was. The good news was that it wasn't between the meter and the stopcock, which meant it wasn't in the neighbouring field, but somewhere in our driveway. John disappeared to his shed and returned with some metal L-shaped rods and walked with them up and down the driveway.

'So another one of your many accomplishments is that you're a water diviner?' I scoffed. I watched him with a smirk until he stopped when the two rods crossed each other in his hands.

'No way!' I exclaimed in amazement.

Both he and the official from the water board smiled.

'My dad taught me how to do it.' John smiled, 'It's actually pretty easy.'

This was my first experience watching someone "dowsing" for water, and I was amazed. Not only could my husband cook, do plumbing and electrical work, but he also had superhuman abilities!

After hearing what it would cost for the water board

to dig up and replace the pipe, John visited our neighbour next door, who rented their cottage from the previous owners of our house, to see if he would ask them where the water pipe came into the house as they had helped build it. John explained to Ron, that for some reason, they never acknowledged us waving or saying hello, and we didn't feel comfortable going to their house to ask them directly.

The next day, Ron sheepishly told us that they were refusing to help us, and he explained the story of our house. After gaining planning permission to build a "farm workers" bungalow next to the original farm, to supposedly accommodate a daughter and her husband who worked on the farm; they moved out and it was rented out as a holiday cottage. They had attempted twice in the past to have the Agricultural Occupancy Condition (AOC) or agricultural tie lifted, but failed and put the house on the market in an attempt to show there was no demand for agricultural dwellings in the area, and could thus remove the AOC by showing that the house had failed to sell. If they could prove the house had been marketed and had no realistic offers from bidders who could fulfil the terms of the tie then they were home and dry. They could sell the house at the full market value without a tie and make a tidy profit! As we had given them a reasonable offer, they felt we had "forced" them to sell the house. Once we knew this, it was clear why they resented us, and why they refused to tell us where the water pipe lay.

Never one to let small things like bitter and twisted neighbours hinder him, John hired a mini digger and, over the next two days dug a long trench from house to stopcock and installed a new water pipe, which cost us a fraction of what the water company wanted to charge. As he had gained his water regulations certificate at college in order to install the central heating, he could officially and legally do this. As an added bonus, he taught our delighted son to drive the digger, and Kieran dug me a hole in the back garden near the house, which within a year was transformed into a decked seating area with a beautiful lily pool.

The sheep had been and come back from their romantic visit to the ram in Devon, hopefully pregnant. Even Spotty; the sheep with the flystrike appeared to have recovered her health and some of her fleece, and had gone along too. We didn't really expect her to come back pregnant, but we took her along for the company. Sheep are social animals and need companionship, and we couldn't have left her in our paddock alone.

We lost our favourite sheep, Snowy, a few weeks before lambing to twin lamb disease, and had to drag the carcase out of the field on a tarpaulin to the yard, to await the specialist firm to come and remove her. Now any experienced farmer will tell you that a sheep is born to die, but of course, you can't tell a beginner this as they will believe that it will never happen to them. It was heartbreaking, a huge loss both financially and as a base for a breeding flock of sheep. We made sure the remaining two ewes were happy and healthy and a week before their due date we moved them into a makeshift pen in the shed to be able to keep a close eye on them. For a week, I visited the shed every two hours, day and night. They just stared at me, the way sheep will. There was no sign of anything. No pawing the ground, no stargazing like the books said, no bag appearing. I was beginning to wonder if I'd got the dates wrong. Then one morning when the alarm went off at five am as usual, I was so tired I decided to turn over and ignore it. When John woke me at seven to wake the kids for school, I wandered out to the shed to find one very happy mum with two lambs. She had obviously waited for peace and quiet and got on with the job without any human audience. The lambs were healthy and suckling mum who was munching contentedly on hay.

The other ewe decided to lamb the next morning, and my first view was of mum's rear with a lamb's head sticking out, but no sign of front hooves! Common sense and reading and re-reading the sheep manual had told me that I must re-position the lamb correctly to ensure it was born safely, and to safeguard the ewe. After a very quick wash, I squirted lube on my hand, slid it inside around the head and realised both legs were back. There was no alternative but to gently push the lamb back a little and squirt lots of lube in, and try and manipulate the legs to come forward.

Trying to "see" with your hands and not your eyes when your head is buried in stinky fleece, and you are totally inexperienced, is all part of the fun; as the lambing instructor told us previously on a one-day lambing course. Luckily, I managed to feel back down the neck to a leg and cupped my hand over the hoof. I bent the leg and managed with some gentle manoeuvring to position the leg underneath the chin. This was enough for the ewe who decided to give a big push and the lamb, assisted by the whole cylinder of lube, slid easily out onto the straw. Mum turned round to lick him dry and clean and my heart soared at being part of this miracle of new life. A second lamb came without any issues, a few minutes later. The children came out to see and were thrilled and John brought me a cup of tea, noting my red face and soiled jeans and realised it hadn't all been plain sailing.

'I'll take the kids to school, and you can go back to bed,' he said smiling, 'although maybe you want a shower first?'

4.
Busy Little Bees

Sadly, in the case of long distance relationships, you are never really sure of the full picture. I'd been noticing "something" different in my mother for some time, but couldn't put my finger on what it was that was concerning me about her. It was very difficult to quantify this over various phone calls as she was so far away. We had never had that wonderful mother-daughter relationship that some people had; in fact we had always had a very cool relationship. She had her lifestyle and I had mine and there were no points of common interest. Mum had been widowed some twenty years or so previously and lived alone in a small upper flat in a Glasgow suburb within a minute of my brother's house.

She had never worked, and had been a stay-at-home mother, bringing up my brother and myself. After my father died she made frequently visits to relations and close friends in the United States; but as she grew older she became less able to cope with the journey and had a couple of health issues which meant she wasn't travelling as much. She then stopped seeing her friends at home as much as before, and had a couple of falls in the winter, which affected her confidence and she became a bit of a recluse. My brother and I both encouraged her to take a taxi to see her friends. The cost was minimal, and keeping her social circle was important, but she characteristically ignored us; hinting that she didn't need her friends. She was particularly derogatory about one in particular, which we found odd at the time, as she didn't elaborate on what had soured the friendship. Then she had another fall on the ice near her home and my brother took to dropping off milk and bread every few days, as she refused to go out at all.

In fact, he told me he saw her most days, as she was increasingly demanding, phoning him at work almost daily for the most ridiculous of reasons, and even though he saw her briefly every day, it never seemed to be enough and she was inventing excuses to make sure he visited. My telephone conversations with her changed slightly, and instead of her asking about the children or the weather or what we were doing, the conversations were from her side, a diatribe about so-called friends. I also noticed more and more references to her "horrible" neighbours and the increasingly important theme of people "spying" on her. Talking on the phone was increasingly uncomfortable, listening to her moaning firstly about lifelong friends and then family in such a way.

At first this was funny; when she referred to her closest friend as a secret spy for the council who had a hidden camera in her coat buttons. At the start I was convinced she was making this up to inject a bit of humour into "duty" phone calls. After all, I could only talk about sheep or fields or chickens and this was very boring to her. But when she repeated the same story every phone call and then, like a child's lie, it grew and grew, and then changed to accusations of theft, I realised something was very wrong. I arranged a flight and went for a weekend visit.

The house was untidier than I'd previously noticed and Mum was not her immaculately-dressed self. I remarked that she needed to make a hair appointment, and she laughed and said she no longer used her favourite hairdresser as he had deliberately burnt her head. She also told me she had stopped going to her regular dentist as he was dirty. However, she was glad to see me, as she wanted to arrange her eightieth birthday party, and announced that she wanted to have it in a small village that was nearly an hour away from her home. Only one of her friends lived there, so I tried to reason with her that it was very remote for her other friends and relations to visit, but she stubbornly insisted and informed me that she had booked the hall and a DJ. And so, despite all the guests remarking negatively on the location the party was set to go ahead some few months later. After sorting out various paperwork and banking issues for her, I was glad to return home, although concerned that she looked frail and was more than a little confused. I remember remarking to my brother on the way to the airport on her increasing sarcasm and nastiness. He laughed and told me we would both be like that when we were old.

Meantime, our smallholding started to take shape. The sheep were thriving and by day they were all allowed out into the paddock, but brought into the shed at night for a few weeks till we were reassured the lambs wouldn't be cold or at risk from predators. We learned not to talk about our livestock much at the pub as we were ridiculed one night for our lack of experience.

'They're SHEEP!' a farmer at the pub shouted from the other end of the bar. ''Ave you not got sheep in Scotland? Or do Scottish sheep live indoors? Get them bleddy out now!'

Sadly, at the pub we had to endure rather a lot of comments about our lack of farming knowledge over the next few years.

A chance conversation online led to me rushing out one afternoon to collect a couple of Aylesbury ducks free to a good home. John sighed and went to start building another bird house, as I popped them in the fenced back paddock. The next day he found an old plastic bath which he sunk in the ground so the ducks would have a swimming place.

We then bought four more young ducks at auction for a few pounds, as the drake was giving the solitary duck no peace at all and we were frightened he would tread her to death. "Treading" is the polite term for poultry sex.

I'll now explain the difference between chickens and ducks having sex for those of you who have never witnessed this. Chickens don't have penetrative sex. The male jumps on the hen's back and with a little flapping of wings manoeuvres himself to spray semen at her vent area. Ducks, on the other hand, do have penetrative sex. The drake has a penis shaped almost like a corkscrew, and mating usually takes place on water as the drake is quite heavy. We didn't actually know this until one day, whilst watching him clamber on and off another duck, we had a full eye-opening experience! We had heard tales about keeping the ducks and the chickens separated for the mind boggling possibility of him taking a fancy to the hens, which is why we supplied him with more concubines. I can imagine the fear in any chicken being subjected to what would only be a highly damaging and painful physical outrage.

Sadly, about a week later, we found one young duck drowned in the bath. We think the drake had surprised her there, had his wicked way, and the weight of him held her under the water too long. We immediately redesigned the bath swimming area to provide a ledge under the water for any ducks to stand on and step out, and prevented Casanova, the drake killing any more of his ladies. Smallholding is all about learning and adapting, but we felt the death keenly.

The pear and apple trees that had been originally planted in the front garden struggled with the poor soil and I finally made the decision to re-plant them in the paddock at the rear of the house. A friend in North Wales had recently begun a small cider making business and was urging me to try the same, but we really didn't have the finances to buy the equipment, but I could plant the trees, which I had been grafting myself from a variety of Cornish and traditional cider varieties. Producing my own trees was not only satisfying but saved money, and I was able to cultivate some rare varieties.

Reading about permaculture expanded my knowledge and to ensure the apples were properly pollinated meant I had to understand more about bees and their importance as part of a symbiotic system, and how they would also contribute to our smallholding. I had joined the local bee-keeping group and soon was the proud owner of two hives, one small colony of bees, assorted equipment, and had the very useful reassurance that my beekeeping mentor was just a mile down the road in Wadebridge. I went to hands-on meetings at the club apiary and learned much more from these meetings than I could ever read in a book.

Bee keepers are a friendly bunch, including a chap from near the South Cornish town of St Austell. We had got to know each other on the internet through a forum called *Downsizer,* and had similar interests. Steve had built his own Top Bar hives and was giving a demonstration at a nearby village, if I wanted to come and see. Now, even though I had Smith hives, I was keen to see these supposedly more bee-friendly hives and arrived at the garden in the village, requiring no directions as the garden and main road was full of people walking around in white bee-suits. There was no nuclear explosion or incident, despite the looks of curiosity from passing drivers.

One of the colonies had literally just swarmed and everyone was trying to see where they had flown to. Normally, when a colony of bees swarm they initially don't go far from the hive, and indeed, they were located very soon in a garden hedge a few hundred metres away. What did surprise me was seeing Steve in his shirtsleeves with no veil or protection of any sort, collecting the swarm in a box and taking it back to a waiting empty hive in the garden!

I was mesmerised as he handled the bees without any fear, even though they were crawling in his hair and beard! He had explained some time previously that because he moved quite carefully and slowly, he gave the bees lot's of time to get used to him being there, and he smelled of the colony. If you get the chance to do any bee keeping, enjoy the smell of bees, honey and wax as you check through the hive. There is also the comforting droning of working bees, who accept you as one of them if you are careful and lift frames carefully and slowly. Steve explained how the design of the top bar hives differs from standard hive design as there are fewer parts to move and the bees are really allowed to live without much human disturbance. He took his time to answer questions and demonstrate checking through the hive, whilst explaining to the poor beekeeper who had seen his bees swarm that he had missed the queen cells on a frame within the brood chamber, and now would have to wait till a new queen hatched. Luckily his original colony was now happily re-housed in the spare hive.

Although I never ventured into top bar hives myself, I learned so much from this visit. With bees, there is no right way or wrong way to keep them. What works for some keepers doesn't for others, but you learn something new about them all the time. They really are fascinating creatures who can work with you to pollinate your garden, fruit trees and crops, and if managed properly will give you honey in return. I had sited my hives next to where the orchard would be, where they would have fruit trees and a hedgerow of native bushes and trees to keep them supplied with nectar and pollen for the whole season they were active.

With the help of friends we finally erected a polytunnel in a sheltered area to the rear of the workshop. As it was a warm, still day, perfect for this technical and demanding operation, we spaced the hoops four feet apart and buried the polythene, as we were terrified of the strength of the Westerly gales that seemed to be a regular late autumn feature in Cornwall. Luckily, all went as planned without any hitches, and we finished in a few hours. Many hands make light work, and you really do need at least four people to cover a polytunnel. Inside we arranged a raised bed along one length and a platform of staging for young plants and a potting table. One ingenious item John managed to incorporate was a washing line strung along the centre of the hoops. With the damp Cornish weather, at least I could dry my laundry in there.

As a later observation, we realised the polytunnel produced an enormous amount of tomatoes, cucumbers, peppers and plants to sell, and also housed our grass cutting tractor. However, we felt we were not really growing the right plants or crops to make the investment worthwhile, and were thinking for ways to supplement this.

The year was on the turn, and autumn brought a small bounty of fruit and vegetables. I picked sloes and made sloe vodka, which would be ready at Christmas, and collected hazelnuts from the hedges around the smallholding. Our Bramley apple tree provided apples for the freezer for cakes, pies and crumbles and Colin brought an enormous bag of plums, which joined them. The freezer was groaning with great produce and thankfully this also saved us a lot of money. We made more cider, this time adding cider apples gifted from a friend in a nearby village.

Our raspberry canes produced a bumper crop, so I made jars and jars of jam. Mum had taught me to make jam years previously and I'd changed her recipe slightly to reduce the amount of sugar.

****

Raspberry Jam

2 lbs fruit
2 lbs sugar
Juice of one lemon
NO WATER

Clean and prepare the fruit (without washing). Put berries in a jam pan and heat on a low heat till the berries release their juice. Cook gently till tender. Now add the lemon juice and the sugar and stir well until dissolved. Bring to the boil and cook until setting point is reached (about 20 minutes). Skim off as much scum as you can. Warm clean jars & lids in a low oven. Pot whilst warm and fix on lids. This makes about 4 lbs of jam. Remember to label with type & date.

****

At the village, one of the regulars had lost his license for drunk driving, and was regularly seen walking home at the end of the night. We later heard the six mile walk was too much when he was blind drunk, and he was often found sleeping in an open shed on the road out of the village. Driving under the influence wasn't uncommon, and another local had to endure gales of laughter at the bar when he appeared one night with masses of scratches to his face, which were as a result of colliding with a hedge whilst cycling home drunk the night before. We began to park the car away from the pub car park.

John worked hard and long hours and the kids had settled into life in Cornwall, but despite our hard work, we were hardly breaking even with the business. This became an obsession with me, as I was terrified the Enforcement Officer would turn up again. In desperation I enrolled on a college course to gain horticultural certificates and more knowledge, hoping this would help our struggling business, or give me more qualifications for a relevant job in horticulture. The hours were part-time and the course ran from September till June, so didn't really affect the nursery business. I have to say, with my experience of horticulture and plants, I didn't really learn much from the course, but the qualification at the end was the big incentive to travelling 45 minutes three times a week down to the agricultural college campus at Rosewarne.

Meanwhile, back in the real world, I managed to enrol us both on a few courses and events at the local agricultural college, subsidised by European Union funding, and this developed our knowledge even more, and also helped us network and make friends with similar smallholdings. This was a great step forward for us, and we suddenly had access to people who could answer all sorts of questions, suggest suppliers and help us out with teaching us different skills with livestock.

Together with my horticultural courses my qualifications and knowledge was adding up, and I decided to enrol on a basic teaching course, suggested by another friend who had previously taught part-time adult education courses in Chester. Spread over ten weeks and only a few hours weekly, I found myself thoroughly enjoying the teaching side of life, and was invited for an interview with the local council Adult Education Department. After a relaxed "chat" and a discussion of what I could offer, I found myself employed to deliver horticultural and basic poultry-husbandry courses.

This really was a major turning point. I thoroughly enjoyed meeting people and sharing knowledge. And the one-day workshops I was offering through adult education were snapped up. If only the nursery was doing as well. We also finally heard back from the estate agent that the family we had bought the house from had withdrawn their offer to sell us additional land. Although annoying, we were not surprised. How people hang onto their hate! We would have to have a plan about what to do with our little flock of six sheep as we didn't have enough land to keep them all. Luckily, I was volunteering at a friend's farm where they had a large flock of 600 sheep and we sadly parted with them. No more sheep for the time being, but I loved the whole sheep and lambing experience and it wouldn't be long before sheep returned to our smallholding.

A large second-hand freezer was installed in the shed, slowly filled with fruit, vegetables, chicken and duck. Another addition to our table and freezer were partridges and pheasant. We had seen a pair of comical, beautiful French Partridge wander in through the front gate soon after our arrival, and commented to Colin; a local who had heard that John was a welder by training, and who brought a bewildering and never-ending array of machinery for John to mend. In addition to the payment for the welding, occasional deposits of a brace of partridge or a pair of pheasants would arrive unannounced; sometimes sitting at the front door or hung in a bag on the gate. People can be very kind, and Colin became another good friend. We were settling into our new life and growing in confidence and experience.

One day, whilst chatting to a friend who wanted some pure breed hatching eggs but was finding good ones difficult to find, I had a brainwave. I could sell hatching eggs via an on-line auction site. I had the pens of pure breed birds and too many eggs, which I was selling for next to nothing to our local shop. Soon I was selling them on-line for a much higher price, and the eggs that failed to sell were sold at the farm gate. As they were fertilised eggs I legally couldn't sell them at the shop. It worked excellently, and finally, keeping chickens became profitable. As we were based so far south in the UK, I could test a few eggs to check fertility, sell much earlier in the spring and get the highest prices. Careful packaging of the eggs meant I received great feedback too.

Our postman, Les, drove in with the usual bills and a small package. We hadn't been expecting anything and instead of the usual lawnmower parts arriving for John, this small heavy package was addressed to me. With no return address, unfamiliar writing and no note inside, I examined the contents on the kitchen table. A small plastic drink bottle filled to the brim with a dark golden liquid.

Kieran quipped that someone had sent me a bottle full of pee! It certainly looked like pee, and John and I stared at it.

'Have you upset someone?' John asked with a smile.

'Not any more so than normal.'

So I opened the bottle whilst the kids and John looked on expectantly.

It smelled...slightly familiar, but I couldn't place it. I lifted the bottle to my lips as Kieran's mouth fell open in horror.

'Mum! NO!'

Too late.

'Not pee,' I announced confidently, after a small sip. How I felt experienced enough to identify the taste of pee was beyond my comprehension, but this was not pee. A moment later, I'd identified the taste.

'It's mead.'

Relief all round the table, but mainly from me. It could have been a very different scenario, and I could have been left with the horrid idea that I'd upset someone locally enough for them to send me a bottle of piss in the post!

Mead is a very old traditional alcoholic drink made

from honey; but I still had no idea who had sent it. That answer didn't materialise until I received an email about a week later asking if I'd received the package. My friend, Gareth, from the Downsizer community, had been following my experiments with cider and sent me a sample to see if I wanted to try mead making! Mystery solved.

Meanwhile in the village, a wedding was announced. A widowed lady, in her golden years, shall we say, was betrothed to a bachelor from another part of Cornwall. The gossip in the pub was bawdy. We, being relatively new to the area ignored the idle gossip until Kate returned from the shop one Saturday full of laughter and news. A delivery driver had come to the shop in fits of laughter telling the owner (with Kate listening) that he'd delivered a parcel to the newly-weds who had answered the door 'In their skin'. Of course, in a small village news travels fast and the whole pub was soon discussing what the "happy couple" could have had delivered. A selection of saucy underwear or marital aids were the most frequently suggested by the regulars at the back bar.

At the other end of the UK, my brother was becoming increasingly exasperated by my mother's behaviour. After having arranged for her to baby-sit for an hour, so he and my sister-in-law could both attend the annual parents' night at the school, she spectacularly failed to turn up. Words were exchanged and tempers lost on both sides. I had to listen to two completely different stories on the phone and attempt to broker a peace despite also being cross at my mother's increasingly selfish attitude and inability to recognise the harm she had done to her relationship with them.

In October the weather worsened and we struggled through gales and lashing rain when closing up the chickens and ducks, and settled in behind closed curtains in front of the wood burner, listening to the wind wailing round the house.

As we were located on the top of a small hill, we were used to the wind from all directions, but this night it was louder than usual, rattling the chimney pots and howling down the chimney. John took Ben, our golden retriever, out at half past ten for his last visit to the toilet and returned looking grim.

'I think we might have problems. There's nothing we can do now, but two of the chicken houses have blown across the paddock to the back wall. It's pitch black and too wild to try and sort them now.'

He had just finished saying this when the lights went out. As he groped his way to the kitchen and found a torch, he told the kids to head for bed. The power stayed off all night, and in the morning we went out to see what damage we had sustained.

The wind had indeed blown two of the largest chicken houses over, but apart from sawdust everywhere, the hens seemed to be fine, and walked out pecking as if nothing had happened. A huge branch had been ripped off the Leylandii tree in the rear paddock and John dragged it off to chop for logs for the future. Thankfully the polytunnel appeared unharmed, which amazed us both. Spacing the hoops closer together and digging in the polythene cover appeared to have given the structure more strength, and we counted ourselves very lucky.

*****

By this time, John had managed to find a welding job at Newquay, about a half hour drive away, and was also teaching welding part-time at the agricultural college. When John had been working in the family business with his father, he had been a professional welder and fabricator. He had wonderful skills and decided to utilise these. He could earn better money and, frankly, our need was greater now as we decided we couldn't throw any more money at the plant nursery as it was never going to be profitable. My courses were seasonal and only for a day at a time, and didn't really amount to the smallest part-time job, and I had applied for another part-time one at The Rural Business School as a Skills Co-ordinator. If I was successful, then I could combine this new job with the teaching events and we would shut down the nursery, which we now reluctantly accepted had failed.

Mum was of course delighted that I was at last getting a 'proper job' and would stop 'playing' with hens. She never understood what I was trying to achieve and really didn't care as long as she could tell relations that, at last I was conforming to her idea of what was a suitable occupation. There was absolutely no point in trying to explain the whole notion of sustainability to her and risking yet another argument.

I quite fancied having an old fashioned clothes airer above the wood burner in the kitchen. In Scotland we call this a pulley. John found some old cast iron ends and bought some timber for slats and a pulley to raise and lower it. This meant we could sell our tumble drier, another step along the path of simpler living, which also saved us about £150 per year in electricity. Yes, the house was looking a little like the Weasley House in the Harry Potter films, but it was also becoming a home. We sometimes had lambs in cardboard boxes in front of the new wood burner in the kitchen, and the dogs slept on the old armchair next to it. In summer, bunches of lavender and herbs were hung from it to dry and the sink doubled up as a duckling bathing station when we wanted to let our incubator-hatched ducklings have a little swim to clean themselves.

The old wood burner in the living room was past its best with a warped door that sucked in air and really needed replacing, and we thought long and hard about what to replace it with. Option one was a straight replacement, whilst option two was a wood burner with a back boiler, which would also run four radiators and give us a tank of hot water. John sat down with the various brochures and then, with a page full of figures and diagrams, did a bit more research on the internet. Within a few days, he had made a decision and explained it to me. Choosing option two would mean running a separate secondary heating system. This was the simplest way and also a good contingency plan if the LPG or electricity cut out. We would have extra radiators in three bedrooms and the bathroom with a heat sink (radiator) in the kitchen. Hot water would be supplied via a third tap at the bath and at the kitchen sink. It would cost about £500 more to install than option 1, and about a month longer to actually fit, but we would be able to recoup the cost within a year, and it would make the house warmer at no extra cost. It would also reduce our reliance on the costly LPG gas, as we could source logs and wood almost free locally.

We carefully removed the old wood burner and installed it in the big agricultural shed, which had been slightly extended. This was originally to provide heat for working inside during the winter. However, Kieran had been pestering John to let him use it as a place where he and his friends could meet and play in the winter, and he wanted to take his electric guitar in there to play it loudly. John said it would be better for the kids to have their own space and as it wasn't in the house or in his way, he was okay with that. There was much whispering and secret smiles between John and Kieran until the suspense was obviously too much for Kieran who blurted out one evening, 'Dad's bought a pool table! And it's in the new shed!' A visit to the shed did indeed reveal a pub-sized pool table with lights installed in the centre of the shed with a couple of old bucket chairs near the wood burner. They had secretly turned my storage shed into a man cave! John did have the grace to smile sheepishly, but said the pool table had come from a nearby pub that was being refurbished and only cost £50.

Times change and our village reflected and marked the changes. Dave's brother, Charlie, died of cancer and there was a grand turn out, with many of the farming community and local gentry attending his funeral and wake, telling tales of Charlie's prowess at Cornish walling and his love of the countryside. The old characters in the village were gradually departing, leaving the village a little depleted of old Cornishmen. However, many new houses were being built and there was local anger at the lack of affordable housing for locals and the influx of strangers. A meeting at the parish hall explained that the population had to rise to keep both the school and the local shop functioning. Too many Cornish villages had lost their schools, pubs and shops and become the seasonal haunts of holiday-home owners. They were empty in winter and ridiculously busy in summer. Although not a universally popular decision, our village had chosen to expand rather than face the loss of the school.

After the funeral I drove up to Ledbury for a professional cider-making course. I had a great time learning the very technical commercial process, and had loads of opportunities to grab advice from other producers on the course, who came from as far away as South Africa, Poland and Denmark. Evenings were spent visiting local producers, and I had a behind-the-scenes tour of Thatcher's Cider. I even met the delightful Tom Oliver, a legend in the cider and perry world and had a very boozy night with Mike Johnson at Ross-on-Wye Cider, trying lots of different batches of cider. It was a wonderful week and provided me with lots to think about as I drove home to prepare to deliver more courses.

I had added Beginners Beekeeping courses to the list of day courses I offered through Adult Education and these proved popular. We had purchased four new bee suits for learners and they enjoyed a four hour "experience" course with a mix of classroom-based information on what was required to keep bees, and the cycle of the year for bees and a long hands-on session working at the hive, gaining confidence lifting the frames up, identifying what they saw - the queen, workers and drones, and how to recognise larva, sealed brood and even watching the emergence of new young bees from the sealed brood cells. There really is something magical about watching new life emerge, and to see the rapt faces of learners who, a few hours previously, had never done anything remotely similar, holding up a finger with a little worker bee happily perched there cleaning her face was wonderfully fulfilling. As autumn arrived, I reflected on what I was offering as courses, and how much I was enjoying my teaching. Our smallholding set-up with our little shed (with the pool table pushed aside) enabled me to share what we were learning with others who wanted to do similar.

I had even written and had published a few articles on smallholding topics for a couple of well known smallholder magazines, which was a great morale booster if rather poorly paid. But I was enthusiastic about how we lived here on our smallholding, and wanted to share all the successes and failures, because they could only help like-minded aspiring smallholders and people loving country life. The tide was turning and finally we were finding our feet financially.

Then, just when you think you have all your ducks in a row, events happen that will mean major changes in your lives.

5.
Snakes and Ladders

November arrived and John stayed in Cornwall to work, and keep an eye on the growing menagerie of animals (we had also adopted a couple of rescue cats by this time), whilst the children and I flew to Glasgow for the birthday party from hell. At least, that was his excuse. Personally I was dreading it, but as a milestone birthday we had to go and endure not just the party, but the build up for the great event.

It was immediately clear to me that my mother was unwell. She had made an attempt to tidy the house, but it was superficial, and I had to clean the bathroom and kitchen before I would use them at all, which irritated her. I also had to secretly bin lots of food from the fridge and do a food shop or we would have starved.

The day of the party was taken up with decorating the venue and she insisted on many balloons and streamers. I cringed at the large banner announcing Happy Birthday and kept to myself the thought that this was all more appropriate to a child's party, but this was what she wanted and it was to be her day.

A glass of wine helped prepare me for the evening ahead and I stoically greeted the small amount of guests who actually made the considerable effort to attend. Despite this, Mum showed a complete lack of empathy to many of her guests. Some of her friends took the opportunity to make telling comments to me about her recent behaviour, and I spent the evening smoothing over ruffled feathers and listening to complaints of her increasingly selfish comments or actions. She herself appeared to be oblivious to all this, and was completely self-absorbed in enjoying "her" night. I tried to talk to my uncle, her brother, about my concerns, but he brushed them off, saying I was imagining it all, and that he thought she was fine, if a little absent-minded. Thankfully, the night went off without any major incidents and the next day we returned to Cornwall, leaving Mum to bask in the memory of her big night.

Autumn was almost over and the honey that had previously been extracted and bottled was sold as quickly as I could get it labelled and the bees put to bed for the winter. The hives were strapped up with cargo straps to prevent badgers pushing them over, and had heavy blocks placed on top. We had noticed deep gouge marks on the front of some of the hives the previous spring, and realised the badgers were trying to access the sweet, golden treasure inside. Hopefully, we would prevent this now. The bees were still flying on sunny days, and the late blackberry flowers and ivy in the hedge provided them with forage to add to the honey I'd left them. We had three colonies to get through the winter and they seemed to be healthy, in large enough numbers to ball together tightly and survive the damp Cornish winter.

John and I made the most of a short, quiet spell and went for a visit to the Lost Gardens of Heligan. We dropped the kids off at school and drove straight there. Heligan is situated on the South coast and is a little warmer than our rugged North coast. The gardens can support more exotic plants and the estate; rejuvenated by the same people who built the Eden Project is a popular attraction for locals and tourists alike. The little orchard at Heligan was still groaning with a variety of local apples, unpicked and unwanted. The late autumn weather was kind and we explored walked and then enjoyed a rare lunch out at a pretty country pub nearby, before returning home to our own little piece of paradise.

I was pleased we seemed to have cracked the secret of working with the seasons and that work on the smallholding was developing into naturally divided slots, with lambing in early spring, lawnmower repairs during the summer and cider making started in September. In winter I was running short courses with Adult Education and we spent the short daylight hours maintaining and repairing our boundaries and machinery. Things were finally on the turn, and we actually had money in the bank.

Putting my mother's slowly unravelling mental state to the back of my mind, I started my new job at the college. Basically, I was to work one day a week, putting together courses with a rural or agricultural theme. They could be practical courses or information- based seminars, and I was soon in my element with courses ranging from chicken health to agricultural planning provision. My hours went up as the courses filled and we had to repeat them. I was delighted because I could do most of the paperwork from home and attend the college campus and other farm for the course delivery where I would register and support those booked on the courses. Together with my other teaching with Adult Education, things were looking up and we became more confident with what we were doing on the smallholding. John was also able to offer day courses on blacksmithing and welding to fill gaps in the college curriculum, also enjoying sharing his knowledge through his teaching. It wasn't horticulture, but technically, we were still running a smallholding with a variety of related work.

Our friends who ran a large sheep farm near Launceston allowed me to help out at lambing time, which greatly increased my lambing knowledge and experience. I loved my evening shifts on the farm, helping with the births, bottle-feeding and making up new clean pens for the ewes and lambs, and although physically tiring, I felt useful, and loved my lone vigils in the barns and sheds. The excitement of trudging down the muddy path in the dark to the lambing shed with my big torch was great. I never knew if it was going to be a quiet night or if there would be lambs popping out all over the place and I'd be run off my feet. But I had quiet moments too, when I'd just sit and stare at expectant mums "stargazing" and making a nest in the straw. Perhaps they would lamb in the next hour, perhaps not. I breathed in the smell of warm hay, of slightly fermented silage, of wool and of little milky lambs, and felt immensely content.

Even when it didn't all go as planned and there were stillborn lambs or hard births, I loved it. Even watching the rats running from their hiding places under the building to snatch at hidden food and then scurry back under cover was diverting. Lambing was a unique and rewarding experience of which most people have never had the chance. The care and attention given to these sheep by their owners was a privilege to see. Lambing time spans between six and eight weeks, and is a physically gruelling time of year. To prevent spiralling costs, most farms enlist the help of family and friends, which is why my offer was useful on this family farm even with my limited knowledge. Despite being exhausted, the farmers took time to visit every sheep and every lamb. Lambs were picked up to feel their tummies to check they had fed, medicating ewes that had endured a long and hard birth, and fostering a motherless lamb onto a mother whose own twin lambs had died at birth.

These men are the salt of the earth, unsung heroes, and I was proud to be of assistance to them. The 600 strong flock soon became a flock of 1400, with ewes having singles, doubles or trebles. As soon as they are strong enough and the weather is fine, the ewes and their lambs are turned out into the fields. This is a sight to gladden any heart and a relief to the farmers that the year and all the effort has worked in their favour, and that they have been winning in the game of snakes and ladders.

Whilst I "played" with sheep as my mother would refer to it, Kate managed to find a Saturday job in the shop and was soon coming home telling me the local village gossip. Every village has its gossip I'm sure, because they are small communities and everyone knows everyone else's business. However, living outside the village, we seemed to only catch up on what was going on through the kids or if it was particularly juicy, or if you were actually there to see it; like the incident at the after school football match one afternoon.

Picture the scene: small boys happily playing football on the pitch, surrounded by their adoring parents calling out encouragement. Or in the case of one mother, calling across the pitch to her very recently-estranged husband, demanding money for new shoes for the kids or something. My friend, who shall remain nameless, and I, stood enthralled. This was better than the telly! I was quickly brought up to speed on the situation before us: *He* had left her for her friend, who also had a son the same age (and in the same class). Very cosy, I mentioned to my friend, who laughed and said the funniest and most inappropriate thing I'd ever heard. 'He's like a dog with two dicks!' and we returned our attention to the domestic drama playing out in front of us, along with the rest of the parents.

The name calling continued from either side of the pitch until the boys, their son and his new girlfriend's son, both stopped playing and stared at them in acute embarrassment. Finally, other parents angrily shouted that this was neither the time nor the place for such a display, and he walked off with the other woman. Who said nothing ever happens in sleepy villages? I reflected that, although it was occasionally inconvenient, we were quite lucky not to live in the village itself. With modern conveniences like the telephone we could, after all connect with anyone we needed to.

Sadly phone conversations between myself and my brother increased as mum's behaviour deteriorated. She had new gas central heating installed to replace the old electric storage heaters and accused one of the plumbers of stealing £100 from her. My heart sank when my brother phoned to tell me.

'It's ok, I've straightened it out, but the firm are insisting I'm there whenever the workmen come to prevent her of accusing them of anything else!'

'I think she has dementia,' I said.

'No, she's at the doctors every bloody week!' he answered. 'I know, because sometimes she asks me to drop her off or to collect her and take her home! She doesn't realise I can't just leave work whenever she clicks her fingers.'

'She phoned me last week', I said, 'asking me to pick her up from the supermarket. I was shocked and reminded her that I was living in Cornwall.'

There was an uncomfortable silence whilst this sunk in.

'Can't you go with her to the doctor's?' I asked.

'As if she's going to agree to that!' he snorted.

'I think I might have to come back up, but I'll need to check I can afford a flight.'

'Look - I'll try and phone the doctor,' he said unconvincingly.

A few days later he phoned to tell me the doctor refused to speak to him, citing 'patient confidentiality'. The only way the doctor could discuss Mum was if one of us was there with her. As he was under pressure at work, he couldn't take the time off 'every time she wants me there'. And then he told me she had called the police a few nights previously to report that her neighbours were breaking into her loft and drilling holes in her ceiling to spy on her. The neighbours had contacted him, fed up and angry with her behaviour, and he had to contact the police to assure them we were 'dealing with it'.

'I need to come up,' I said. 'I'll look at flights and message you.'

John was nearing the completion of changing our attached double garage into a self-contained studio holiday let, which we knew would significantly boost our income and allow us to possibly treat the children and ourselves to a proper holiday. He jokingly mentioned that it was just as well we had decided to make it disabled-friendly. I looked at him in horror at what he was implying.

'No way is she moving in here! We'd kill each other!' I said, throwing clothes into an overnight bag.

*****

When I arrived at my brother's house, we discussed the plan of attack together before both going round to see her. We agreed I would go with her to the scheduled doctor's appointment the next day, and to getting a referral to a dementia specialist, and also to discuss with her the thorny issue of Power of Attorney. I had never seen my mother so angry, when we told her. Even when confronted with a double kitchen cupboard stuffed with bags of stock-piled medication that she wasn't taking she refused to accept she had a problem. When shown rotten food in the refrigerator that had use-by dates from the previous year, or final reminder utility bills that she refused to pay she walked out the room, telling us to mind our own business.

I managed to go with her to the doctor, who looked relieved to see me; stating that she had been concerned at Mum's health and behaviour for some time, as she was making appointments to attend the surgery almost twice weekly for no apparent reason. Patient confidentiality was again cited by the defensive doctor, who made a referral to a dementia consultant there and then. The other thing I had to consider was obtaining Power of Attorney to deal with the looming issue of protecting and helping Mum to manage the mounting paperwork issues she was now finding hard to deal with.

I phoned Jim, my friendly Scottish solicitor, explained the situation and asked about Power of Attorney.

'Well, yes, I can do it,' he said, 'but she has to agree, and I have to judge how mentally fit she is. If she's too far gone, then it's really too late, I'm afraid.'

'Let's make a morning appointment; she's brighter in the mornings at the moment.'

'How….advanced do you think she is?' he asked gently.

'Oh, on some days, she's totally barking,' I admitted before putting the phone down.

So, the appointment came, and after a lot of verbal abuse, ending with my brother threatening that if she didn't go with me to the solicitor's he would stop her visiting her beloved grand-daughters, she finally tearfully agreed. We were having to play hardball and it was emotional, exhausting and horrible for everyone concerned. The solicitor took her into his office alone, whilst I waited, wondering what fabulous stories she would tell him. Eventually the door opened and Jim brought her back to the waiting area.

What a charming and interesting lady your mother is,' Jim said loudly for her benefit, ushering us to the door. 'I'll get the paperwork sorted for you and be in touch soon. It was a pleasure to meet you,' he said to Mum as he shook her hand.

He looked at me over her head, and whispered the words 'Just in time, I think.'

The visit to the Dementia clinic didn't go nearly as well.

Mum decided not to get out of bed and dressed until I lost my temper and shouted at her. In fact, there was a lot of shouting on both sides. Thankfully, I told her the appointment was earlier than it was, and we made it in time.

The consultant spoke to us both and asked Mum to complete a written test to judge a baseline of where her cognisance was. Mum was angrily silent. She refused to complete the test, despite the doctor indicating that this was her opportunity to show us how capable she was. The consultant explained that there was medication available that would slow the condition down and the test was required in order to adjust the prescription. Eventually Mum began the written test but, after a good few minutes trying to understand a question, threw the pen down in a show of temper.

Within an hour we had been told that, yes, it appeared Mum did have dementia, probably Alzheimer's Disease, and medication was to start immediately. Nothing could reverse the damage the disease had already done, but medication would slow the decline in most cases. She also told us both that we also needed to immediately consider either residential care or supported living, as it was clear from Mum's mental state and unpredictable behaviour she could no longer live unsupported. The consultant was pleased we had organised Power of Attorney whilst Mum sat rigid with anger, and stated there was nothing wrong with her. The consultant told me this was a common response and that she would see Mum on a monthly basis to monitor the medication and the illness. Whilst Mum was in indignant denial, my head span with the diagnosis and the implications for future care.

After explaining the situation to my brother, I flew back to Cornwall to discuss the situation with John, and the implications of what exactly "supported living" would mean for us all.

Less than a week later, Mum had deliberately flooded her

downstairs neighbour by putting the plug in the bathroom sink, blocking the overflow, and leaving the taps running whilst she went off for the day to meet a friend in Glasgow. When confronted on her return that evening by my now-very-angry brother, who had been summoned by understandably irate neighbours, she imperiously commented that the neighbour was a smoker, and the smoke was coming into her house through the sink. The next day we discovered from the insurance company that Mum had cancelled the house insurance some six months previously. Thankfully, we were able to put this back in place and pay the unpaid electric and telephone bills she had ignored and I arranged direct debits to ensure this would not happen again.

It is a horrible thing to see a previously witty and clever and fiercely independent person robbed of the ability to look after themselves. Yes, as we grow older we expect to need a little help along the way and accept old age as part of a natural process, but to lose your independence and friendships because of a disease that robs you of the very things that shape your personality is cruel. We discovered from other neighbours that Mum had been seen at the local supermarket begging money to pay for her shopping from strangers because she had left her purse at home, and that she had more than once allowed strangers into her house.

A series of phone conversations between me, the consultant and my brother resolved that she was now a danger to herself and others, and that the immediate choices were residential care or moving in with a relative. This lead to more than a few "helpful" relatives suddenly pointing out my *duty as a daughter* to take Mum into our home permanently. Interestingly, none of these well-meaning relatives offered to have her themselves. The supposed close-knit Armenian family system who looked after their own didn't exist, apart from my poor uncle, who was in shock, unable to offer any practical help.

'What can I do?' I said to John one evening. 'She's not mad enough to go into care, but can't be left to live in her own house. I don't want her here, but there isn't anywhere else for her to go unless it's a care home, and she's not bad enough for that yet. I feel as if the ground is slipping away from under my feet, that we have no choice but to take her. And it's going to be awful.'

John was incredibly calm and supportive. 'What else can you do? She's your mother, and needs your help, even if we all hate the idea. She's not ready to go into a home, but who knows how long it will be before we have to think about that.' He was the voice of calm when my world was collapsing around me.

'Who knows?" he added. 'She's over eighty now. Maybe she'll die soon and it won't be a problem for too long.'

Rather than worry about something I had no control over, I tried to push all this to the back of my mind and as it was May, I collected lots of elderflower blossom and made wine and cordial. The season is so short and you have to collect the fully open elderflowers on a warm day. Some trees produce flowers which smell a little like cat pee, so choose your tree carefully. It's a great excuse to wander round the lanes in sunshine, collecting full flat elderflower heads in a basket and listening to the bees in the blossoms delightfully announcing that summer is on its way. And of course, the resulting cordial can be enjoyed all year to bring back summer memories.

****

Elderflower Cordial

25 Elderflower heads
1 kg sugar
Juice & zest of 3 lemons
1 tsp of citric acid(optional)

Place the shaken heads into a fermentation bucket with the lemon juice & zest. Pour on 1.5 litres of hot water. Cover and leave overnight. Strain through muslin or a very fine sieve into a jam pan. Add the sugar and the citric acid, heating to dissolve the sugar. Simmer gently for a few minutes then pour hot into sterile bottles. Add a pinch of a crushed Camden tablet if you wish to keep it, or pour into plastic bottles and freeze. This is lovely in ice-cubes (with perhaps a few individual blossoms added) for a delicious Gin & Tonic or for children's drinks.

****

John's business grew with more and more people bringing garden and agricultural machinery for repair and servicing and finally we were bringing money into the holding. I reinvested money into the provision of juicing equipment following two cider-making courses that were run at our smallholding via The Rural Business School. I also drove to North Wales for a weekend to stay with my friend John, who had started his own small cider-making business and came back enthused and ready to begin. This was a good distraction from the whole nightmare world of my mother and although hesitant to go, John persuaded me to take the opportunity as we didn't know what was round the corner.

Unlike hardy plants, people always seemed willing to buy alcohol, and indeed, the whole artisan cider business was enjoying a U.K. wide resurgence. Unlike beer making, small-scale cider makers also benefited from a duty exemption as long as production stayed under 7 hectolitres annually (7000 litres). As my friend had repeatedly told me, it was too good an opportunity to ignore.

Whilst I was away, my fabulously clever husband used his skills as a fabricator and welder to build me a rack and cloth press that dismantled if we needed to transport it. This saved a considerable amount of money, and we just had to wait for the harvest. Not that we were bored on the smallholding. There didn't seem to be enough hours in the day to keep up with nature.

Our lovely bees, continuing daily life oblivious to our personal problems, decided to swarm one afternoon, and we quickly sorted out a new clean, empty hive for them, having seen where they flew off to. They had all clustered at the top of a thirty-foot tall Leylandii tree in the back paddock. John fetched a big extending ladder out ready and we assembled the tools we needed: a feed sack, secateurs, loppers and a soft hand brush. As I have a thing about ladders, John was the swarm catcher for the day. The plan was simple: ascend the ladder, use the loppers to trim the branches around the swarm, cut the branch and catch the swarm in the feed sack, return to the ground. What could go wrong?

We stood looking up into the six metre tall tree.

'It's a long way up,' I said.

'Ten minutes and I'll be done,' he said, already sweating in his bee-suit, and climbed up the ladder.

I stood back, mainly to see better what he was doing, but able to run like mad if the bees got spooked, because I wasn't wearing a bee-suit.

Well, ascending the ladder went well. John tucked the feed sack between his legs and started to remove the impeding branches to expose the swarm. The trouble was that the tree was very broad and, even with his weight pushing the ladder into the tree, he couldn't quite reach the swarm. Eventually he caught the slim delicate end of green new growth at the end of the correct branch with outstretched fingers and started to gently pull it towards him, inch by inch. Almost there. A little nearer. The tension in the branch was almost palpable where I was watching from the ground. Finally he could carefully manoeuvre the loppers to cut the branch and take the swarm.

Then the soft green growth snapped with a twang and the branch ricocheted back to where it had been, releasing a cloud of angry bees noisily into the air. Always supportive of my husband, I ran from the paddock to a safe distance. John slid rapidly down the ladder and backed off from the angry bees, now trying to reform the cluster around the queen on the branch. Some few minutes later, most of the bees were again all formed like a tight brown rugby ball around the branch. John's second attempt to collect the swarm went like clockwork and he brought down the feed sack complete with the branch and the swarm, and we carefully installed them in their new hive. Then we left them to settle down, whilst John removed his sweat-soaked bee-suit and clothes and took a shower. All being well, this would be our third colony of bees, and we needed to make up a spare brood box in case we needed it in the future.

This was great as a colony of bees alone without the hive cost at this time around £150. That was for a small colony, called a *nuc* or nucleus, with a queen and around 5 frames of brood and some workers. The colony would quickly expand over the summer and be a welcome addition in the orchard.

Finally, I coaxed Mum into coming for what I explained

to her was a "holiday", just until she was back on her feet again. I had already arranged with the GP and consultant to transfer her notes to Cornwall, and bought plane tickets. After a glass or two of wine at Glasgow Airport, I manoeuvred her onto a plane and she was installed in our newly completed annexe, 'For a holiday until you are feeling better,' we told her. Again.

I was beginning to see that I could have the same conversation with her over and over before it registered. John drove to Glasgow, collected her furniture and belongings and I began the hardest and frankly, the most thankless episode of my life. So much for the anticipated additional income from holiday lettings.

Although the children loved their grandmother, this was at a distance, with visits, and when grandma was "normal". The reality of her living with us was very different. I tried to explain it as best I could.

'When I was little and needed someone to look after me, she was there. It's the least I can do now she needs help.'

'But why do *we* need to do it?' Kate whined, understandably upset.

'Because no-one else will have her,' I said simply.

She was visited soon after her arrival by my uncle and his wife. We showed him her beautiful annexe with its own kitchen area, sitting room with television, her own phone and the disabled-friendly shower room. John suggested a trendy local bistro and we happily went out for an extended family lunch. There was a little incident when Mum completely forgot what she ordered, and when her fish arrived, she complained loudly that this was not what she had ordered, but my uncle smoothed it over and offered to swap his steak pie for her salmon, which she then again changed her mind about. We returned to the house for coffee, and I remember mum begging him to take her with them, and telling him we had made her a prisoner.

'Hardly a prisoner,' I said crossly.

'But she just wants to go back to her own house,' my uncle pleaded. 'Why can't she go home?'

'What? Flood the neighbours again, phone the police and not take her medication? Accuse everyone of stealing from her, and have a fridge full of two-year-old food?'

This was all a terrible shock to him of course; she had told him nothing of her recent actions. He really believed she was mentally okay, a bit forgetful and spiteful, but essentially okay. This was part of Mum's problem; she could appear to strangers to be perfectly in control and rational for a short period of time, but she couldn't keep that up for long. Then, as the illness progressed, it reduced her inhibitions and a nastiness that had always been there, but masked, was becoming more prevalent, despite the medication. I felt so sorry for my poor uncle. In the course of one short visit, he had lost the sister he had loved and the shock had aged him visibly. In the back of his head I'm sure he was also considering the strong possibility as a family member that he too might develop dementia, and may end up in a similar situation himself.

6.
Keeping Mum

Late summer was beautiful, with long sunny days, and teenage ducklings living their comical lives in the paddock and pond. However, they were bred for a purpose and had now reached the optimum time for processing. Fourteen were destined for the freezer, and having practised on one, I knew the quickest method to prepare them was simply to skin the birds, take off the breasts, legs and thighs. So, when they went into their shed that night and had settled down, I took them quietly, one by one to the large shed to kill and prepare for the freezer. John never became involved with this side of smallholding life. He couldn't bear to kill the animals, especially if we'd had them from babies; but he was realistic and accepted this is how we lived, the animals all had good lives and quick, stress free deaths. There was no tortuous trip to an abattoir, no stress, just quick, clean deaths after a carefree, fun, free-range life.

Whilst I tried to combine smallholding life with teaching, running the house, keeping an eye on Mum and ensuring she ate and took her medication, John escaped to work in his shed. When it all became more than a little fraught, I occasionally escaped to college for peace and quiet, to catch up on work I could not do at home. Luckily, my boss was extremely understanding, and as long as I was delivering courses, all was well. I could do lots of the work from home and a wide variety of courses were filling and successful. We juggled like mad with work and children and getting Mum to doctors' appointments, but managed. At the beginning.

The honeymoon period didn't last long and Mum began to interfere with parenting. Both John and I resented this and when she started to undermine our authority regarding things like treats, homework and behaviour, we had to reinforce to her that *we* were the adults in charge. Naturally, she resented this and demanded to go home. Despite us both explaining why this was no longer an option and that we all had to try and make the best of a horrid situation, she was living in her own bubble, increasingly defiant. The incidence of inappropriate conversations and comments also increased. The first time this happened outside the house was quite funny. At least, it would have been if you were watching it on TV as part of someone else's experience.

I took her to collect my son from school, and introduced her to my friend who was collecting her daughter. The children erupted out of the building, followed by the teacher, and the playground was full of noise and colour. Kieran happily ran to us.

'Is that your teacher?' my mother said to my son, indicating to a lady standing at the school doorway. 'Do you like her?'

'Yeah, she's nice,' he answered with a friend pulling at his sleeve to play.

'Do you like her because she's got big boobs?'

'Oh my God!' I spluttered, pulling her and my horrified son towards the car. 'Time to go!'

My friend Jane, who had been standing chatting with us, called sympathetically, 'I'll phone you later.'

At bed time, a small voice called from his bedroom.

'Mum, please don't bring Gran to school again.'

'Don't worry, I won't!'

Kate was in his room like a rocket to find out the gossip,

and a few minutes later the silence was filled with much laughing and chatter. At least they could laugh about it.

Determined to make the most of the good weather, we combined beach rugby training with a family picnic. The car was stuffed with windbreaks, a couple of garden chairs, a disposable barbeque and food and drink. The highlight of the day had to be watching Mum tuck her skirt into her knickers to wade out to her knees in the sea. The inevitable happened and she lost her footing and landed on her backside in the water. I commented to John that whilst I had frequently done it when the kids were young, I didn't think I'd have to bring with me a change of clothes for my mother! She sat on a towel all the way home, but managed to see the funny side of it.

One night when John was out at a pool tournament, Kate, Mum and I were sitting watching television, when Mum came out with some extraordinary and horrific story about her mother's death. It came out of the blue, and Kate and I were shocked at both the timing and the nature of the revelation. As Mum had only been four years old when her mother died tragically, I was surprised at the details and wondered if this was some fantasy that she had conjured or embroidered; but I was more horrified that she chose this moment to announce it in front of a youngster. I told Kate to go off to bed and tried to explain to Mum that she couldn't just blurt out such a personal and explicit story like that in front of a child. She had drunk a glass or two of red wine, but I wasn't convinced this was drink talking. I'd become increasingly worried that she needed her medication increased, as I thought she was losing the ability to control her delusions and fantastical story telling.

After seeing her to bed, I visited Kate who was in Kieran's room telling him what had happened. I explained to them both that Gran's illness meant that sometimes her recollection of events were not true or real and I apologised for the incident.

Kieran asked me if I was going to get dementia too.

'I hope not, but none of us know. There's no way to say who'll get it. Look at Grandpa Andy - he's the same age as Gran and he's fine.'

I took the opportunity to broach something that I'd been thinking for some time.

'If I *do* get it, please don't do what I did and have me move into your home and family. It's been very hard and I don't want you to grow ill looking after me or Dad. Look at the way it's affected us.'

What I really meant was that if I went half as demented as her I wished they'd give me some pills or push me off a cliff or something. On really bad days I had deliciously terrible fantasies about killing Mum, which made me feel instantly ashamed and guilty, but they just kept popping into my head.

So it was no surprise to me when I quipped, 'If I go really ga-ga give me a big bottle of gin and when I'm really drunk and unconscious, smother me with a pillow.'

Kieran started to cry and said, 'Oh, Mum! - I could never do that to you.'

I cuddled him fiercely.

'Don't worry,' said Kate. 'I can do it.'

'Yes, but you have to wait till I'm really ga-ga Kate!' I said laughing, and they laughed too.

A few days later, what was to have been a routine visit with Mum to the optician led in turn to a visit to an ophthalmologist who confirmed that Mum was suffering from cataracts, but also from macular degeneration. He tried to explain the complexity of her diagnosis, how her eyesight would gradually diminish, and as she was losing the ability to understand anything she could actually read, there was nothing they could do for her. I think she only understood the last part. Her temper flared and she again denied there was anything wrong with her. The consultant took a menu from the pile of papers on his desk and pointed to one line on it.

'Read that line to me, please.' he asked her gently.

'Roast chicken with mushrooms!' she announced triumphantly, after a delay of about a minute.

'And explain to me what roast chicken is?' he continued. 'Just describe it.'

Silence.

Ever protective, I urged, 'Roast chicken, Mum - what is a roast chicken?'

She handed the menu back crossly. 'I don't know.'

He explained to us both that the disease had interfered with the connections in the brain. This wasn't just memory loss; this was the result of damage to the brain which was preventing the normal processes of understanding and processing. It explained why she couldn't follow a movie, read and understand a book or follow a knitting pattern.

The consultant completed the paperwork and led us to the door.

'I know it seems harsh, but in a few minutes she won't even remember this conversation. Sadly, I see far too many cases like this now. Try to be kind to yourself; you're just at the start of a horrible journey.'

The kids both stopped having friends round and were visibly embarrassed around her. Hard enough bringing up children, but also trying to manage an elderly, feisty dementia sufferer at the same time was hell. I'd like to be able to say we made the best of a bad situation, and discovered more in common with each other as time went on; but I won't lie. It started badly, and continued badly for nine months. The medication wasn't really controlling her behaviour and she was soon prescribed high doses of an anti-psychotic drug to stem the increasing delusions that people were spying on her, stealing from her and trying to poison her. She refused to wash or shower and we had many stand-offs.

One day John could hear the daily argument between us in her bedroom. I was pleading for her to take a shower before dressing. She had been wearing the same clothes for a few days and smelled particularly ripe. I wondered if she was undressing at all or sleeping with her clothes on. John stormed into her room.

'If you won't let Lorraine help you in the shower, I'm going to drag you into the garden and hose you down. You stink!'

'You wouldn't dare!' she screamed at him.

'Just try me!' he yelled and pointed to the shower room. 'Get in!'

Finally with tears of rage, she agreed. It really was like dealing with a toddler, I thought removing her clothes to the washing machine. A half hour later, dressed in clean clothes, it was as if the incident had never happened, and she and John sat down chatting, whilst having a cup of tea. I was living in a surreal universe.

We had good times too and had gotten into the habit of taking Mum and the kids to the local campsite bar a mile down the road from us. It had a bar with a pool table and John, Kieran and Kate could play pool, feed the juke-box and relax while Mum sat at the bar and held court. Thankfully, the owners and staff knew about her dementia and were very forgiving. This was especially important when it came to serving her gin.

'I don't want any of that cheap supermarket rubbish!' she would shout to Pete, behind the bar.

'I'd never *dream* of serving you that!' Pete would retaliate dramatically. 'I know it has to be Gordon's for you!'

After a few large ones, we would manage to get her back into the car and drive her home.

I don't know why we have to leave so early!' she'd protest every Saturday night on leaving there.

'Because it's late for the kids and you've had enough.' I'd say (every Saturday night).

'I can hold my drink,' she would retort.

Like a broken record; the same conversation every Saturday. I was starring in the movie *Groundhog Day*. But at least with a lot of gin inside her we knew she would sleep all night and that would let us sleep all night too.

Some friends were having a combined birthday and anniversary party and invited us all, including Mum. There was to be a marquee with a band and a hog-roast and it was the event of the season in the village. Ever the party animal, Mum graciously overlooked the rustic straw bale seating in the marquee. After a couple of glasses of wine she got up onto the wooden dance floor and danced to the music whilst we sat and watched. The children met up with school friends and were secretly drinking and dancing, but we turned a blind eye as this was an opportunity for them to relax a little. As the evening wore on and Mum's repeated visits to the bar meant she was more and more plastered, we tried to make excuses to go home. Eventually, around midnight we rounded up the kids and Mum and headed back to the car, thankfully parked nearby. Mum tunelessly sang a pop song as she staggered along and the children laughed at her, but the evening had gone well and we had all enjoyed it. The next day she complained about sore hips and stayed in her room, denying all suggestions that she might have a hangover.

Mum enjoyed her trips into town and a visit to the bank, although she had no bills to pay or need for money. She said she liked to feel the notes in her purse. God knows what she did with the money because she 'needed' to go to the bank at least every week to get more. One day she shouted aloud to passers-by right outside the bank in the middle of a busy pedestrian area that I was stealing from her. Thankfully one member of the horrified bank staff ran out and ushered us inside, assuring Mum I wasn't stealing her money.

The accusations were embarrassing and hurtful and made any social interaction very difficult. She made us miserable wherever we took her, either by her words or actions, and soon our social life shrunk to nothing. We were no longer able to take her for Sunday lunch or for an evening at the pub, which she had always enjoyed, which meant we couldn't go either, as we were by this time frightened to leave her in the house alone. Her behaviour and comments were increasingly anti-social. Relations never visited, and they phoned me, instead of her, for occasional updates, and of course, to "advise" me.

Looking back at this time is very hard. John and I had to juggle work, with one of us staying at home to keep an eye on her. This meant our business went downhill and we were losing money. Again. John's lawnmower business was really lucrative, so for four days a week, he would work all hours in his shed, whilst I attempted to do paperwork and course material on the computer, trying to keep an eye on Mum. On days when I had courses, he would do jobs around the house or clean or try to cook without interference.

Mum seemed to have no idea of all the changes we were making to accommodate her and was oblivious to it all. She continued to "advise" me and begged me to 'get a proper job' instead of playing with chickens all day. In my head I laughed as I honestly couldn't remember the last time I'd played with a chicken or spent time idly watching the ducks or the lambs or just stood in the orchard enjoying the peace and quiet. Our relationship with the children dissolved into resentment because we never got any quality private time with them any more. How we managed to hold our own relationship together I will never know; although I have to say John was a tower of strength and understanding. There were frequent dark days where I just stood in the paddock crying with the sheer unfairness of it all.

As Mum revelled in old movies, we invested in a few of her favourites. Old Deborah Kerr and Doris Day films in the main, but I also tried what was then a new release: - *The King's Speech*. The kids decided to sit with us to watch it and we settled down in front of the wood burner with chocolate and crisps and tried to make a family night of it.

'Who's that?' Mum pointed at the screen.

'It's George VI,' said Kate.

'No it's not. It doesn't look anything like him!' she snorted.

'It's an *actor,* Gran,' Kieran explained.

'Doesn't look anything like the King.'

A few minutes later, 'Who's that?'

It was going to be one of those nights and Kate announced she was going to bed, stood up and left, making eyes at me. *So much for a nice family night by the telly*, I thought.

'The Queen Mother. That's Queen Elizabeth's mum,' I explained.

'No, it's not!'

'It's a *film*, Mum. They're actors!'

Soon, it was just Mum and I watching the film. At least, she wasn't really following it as she kept asking every few minutes who everybody was, and I sat with silent tears falling as I reflected on how my family was falling apart.

One Saturday, she had been in her own annexe for some time. This, although a relief was unusual so I casually asked the kids when they had last seen her. They stared at me and replied that it had been at least an hour. I knocked on her door and getting no reply, went in. The suite was empty. So, John and I had to don coats and go searching. I took the car and drove up the lane towards the village, whilst

John headed down towards the busy A39. Surely she hadn't intended to walk to Wadebridge which was two miles along a very busy country road? She didn't even really know where she was. I recalled an interesting conversation with her a few days previously when she told me I could take her back to Glasgow, as it was only half an hour away in the car. I used a map of the British Isles and pointed to Glasgow and Cornwall, in an attempt to prove the vast distance between them, but quickly realised I was wasting my time and she just could not understand. My anxiety levels rose with the picture in my head of a confused elderly old biddy wandering around country lanes with no clue of where she was, or even the address of where she now lived.

Soon, I found her walking along the lane, about a half mile from the house. It was cold and raining a sleety, nasty rain. She had no coat on and was soaking wet. I bundled her into the car and headed home, honking the horn so John would hear me and return to the house. I had to strip her and get her into a warm shower whilst she ranted incoherently. John came through and locked and removed the key from her outside door.

Now, she was unable to leave the annexe through her own front door, and had to come through the house, and by her own actions had reduced her own independence. A few hours later, when she seemed slightly more lucid, she told us that she had been going to the bus stop to catch a bus to Glasgow. Some very angry, tired part of me whispered in my head that we should have left her outside.

Try to remember, the Dementia advisor told me as I sat crying in her office; that it's the disease that you hate *not* your mother. By then I had been prescribed steroids to control irregular heart palpitations brought on by the stress and exhaustion of her living with us. I knew the dementia was an illness and not her fault, but there was no end in sight.

Frankly, at that time, we were looking at the prospect of her being alive for the next ten years, and us having to manage her. I just wished she'd die.

One sunny afternoon, whilst I was on the phone to a colleague at the Rural Business School, I absently watched Mum in the garden from the kitchen window. She was wandering along the hedge slowly. John appeared in the kitchen and put the kettle on to make tea. I couldn't really see what Mum was up to and was trying to explain on the phone about course numbers, when I saw her pop something in her mouth. The only thing that had berries on at that time of year was poisonous.

With my free hand I opened the kitchen window and called out, 'Don't put that in your mouth!'

John came to look.

'Mum! Don't eat the berries!'

Confusion on the other end of the phone.

'No, sorry, Sue, not you. I'm shouting at my mum. I'll call you back.'

John moved closer to the window.

'Eat the berries!' he called out, smirking at me.

'Mum, don't eat the berries!' I shouted, pushing him away. 'It's not funny and you're no help!'

Later that night when we went to bed, she wandered in at midnight, asking if we'd like a cup of tea. I guided her back to her own room, explaining it was midnight and wearily went back to bed. John had the duvet over his head, but I distinctly heard him mutter that we should have let her eat berries all afternoon long, and maybe we could introduce her to mushrooms if that didn't work.

I'd reluctantly been forced to resign my part-time teaching job with Adult Education in order to remain at home with Mum whilst John worked, but was determined to continue my Skills Advisor job, which I adored and which gave me something to look forward to. I was putting together really diverse courses and meeting some wonderful smallholders from all areas of Cornwall who provided training for courses as diverse as horse-logging, first aid, butchery and agricultural planning law. I was learning loads and the pay was good, but the distraction and social interaction were now increasingly important for my sanity.

The Skills Programme was funded by the European Union, and was one of the best things to happen to Cornwall as an impoverished area. There was lots of funding available for training and events to try and encourage farmers and smallholders to diversify and learn new skills, hopefully to improve their profitability and encourage employment within agriculture. The range of experts within the team at The Rural Business School was vast, and I'm proud to have been a small part of this, and have to say it was the best job I ever had.

John was now totally self-employed at home, servicing garden machinery and teaching a few day courses in winter on welding and blacksmithing, which was generating good income. However, our home situation was worsening and I have to say the children had an awful time of it. The atmosphere and the unpredictability of Mum's moods led them to resent me for arranging for my mother to live with us. I was incredibly tired and stressed and had little time to sit and chat with either of them. Being a teenager is hard, but being a teenager when your mother is consumed with taking care of a dementia sufferer is harder still. We no longer went out for daytrips or meals as we couldn't leave Mum alone, and couldn't take her with us as she was rude to everyone. The children were too young to understand that family responsibility means caring for family members of all ages, and school hours shielded them from the constant drudgery of Mum's behaviour and our increasing irritability. We met more and more people who sadly, had the same situation, and had to endure all our friends and acquaintances saying, 'You're very brave.' We didn't feel brave, we felt under siege.

One night she eventually went off to bed in her own room, and we waited for twenty minutes, as usual, till we were sure she wasn't going to reappear to wish us "Good Morning" as had happened so often before. It had been a particularly trying evening with her, and John and I went exhausted to bed.

At 2am we were awoken by the sound of a smoke alarm blaring, and the dogs barking. Quickly jumping out of bed and putting the lights on, we hurried to the kitchen. In the smoke-filled room, she was standing in the dark in her nightie with a knife in her hand. As I flicked the light switch on I could see the toaster spouting flames up the wall.

'What are you *doing*?' I shouted at her, through the noise of the alarm. She merely stared blankly at me. I don't think she even recognised me at that moment.

John quickly unplugged the toaster, which was on fire, ran to the front door and flung it outside.

The children were crying and screaming in the hall, and we opened all the doors and windows to let out the smoke.

'I was just making some toast,' she replied feebly, totally unaware of the chaos around her.

There was an enormous scorch mark and soot all up the wall and on the ceiling from where the burning toaster had left evidence.

'What were you doing with the knife?' I cried, pointing to the kitchen knife.

'The toast was burning,' she answered making stabbing motions with the knife.

'You tried to get the burning toast out with a *knife*?' John shouted at her, realising we could have been dealing with an electrocution. The smoke was finally clearing but the alarm was still deafening. It was like being in some horrible nightmare.

'Why were you making toast in the dark, Mum?' I cried, remembering the kitchen had been in darkness when we rushed in.

'People were looking in the window,' she answered.

John swore and herded the children to their rooms, shouting at the dogs to be quiet. As the smoke alarm finally stopped, I could hear Kate wailing, 'She's going to burn us all alive in our beds!'

About an hour later, John wearily returned after finally getting the children back to bed and closing all the doors and windows. He gave me a hard stare and muttered something about going back to bed, leaving me to try to settle Mum and return her to her own bedroom.

'I'm hungry', she said. 'I'll make some toast.'

I marched to a cupboard, found and thrust a packet of biscuits at her.

'Just go to bed, Mum. Right now! We'll talk about this in the morning.'

'But what about my toast?'

'Just go to fucking bed! Right now!'

I'm sure none of us slept for the remainder of the night. John lay angry and still and I lay with my heart banging in my chest all night, and tears streaming down my face. The next morning was very surreal. We got up as usual to wake the kids, feed the animals, and ate breakfast in silence with the acrid smell of an electrical fire all around us. Kate told me in no uncertain terms that we really had to get Gran into care. I agreed and told her I was phoning the emergency social worker to arrange this.

'Honestly, Mum, why can't we be just a normal family. Just for once!'

This struck home like an arrow and I reeled inside with hurt and a whole lot of other emotions. My world was unravelling, my family falling apart. Meanwhile, John read the emotion on my face, picked up the car keys and took the kids off to school. I had a good cry in the bedroom and when he returned to start washing down the soot-covered walls and ceiling, I phoned the social worker.

His response to our story was that he would see if we might get respite care for the next week or so, but he couldn't promise anything. I took a deep breath.

'Listen very carefully to me. She set fire to our kitchen at 2am this morning whilst we were in our beds sleeping. I have two children living here. Are you *really* telling me you are refusing to find somewhere for her today? Because if that's the case, then I'll drive her to the hospital and leave her there with a fucking note pinned onto her!'

'Now, Mrs Turnbull, I understand you're upset....'

'Upset? *Upset*! You've never *seen* me upset. Has *your* mother ever set *your* house on fire? Are you going to be responsible when my children are burnt to death in their home? If someone isn't here within the hour I swear to God, I'm taking her to the hospital, and I will *leave* her there!'

I slammed the phone down, shaking.

My mother wandered into the kitchen in her nightgown, and seeing John on the ladder, wished him a good morning, and remarked with a chuckle that he was starting work early. Rigid with anger, he stopped cleaning. From his perch on the ladder I heard him clearly shout, 'Fuck off!' It was the only time I'd heard him raise his voice at my mother.

Two social workers arrived before lunchtime, and had a conversation firstly with me, mainly crying, and then with Mum. She couldn't remember anything of the incident, and was angrily adamant that I was lying. It took little less than twenty minutes to have a Deprivation of Liberty form completed for Mum, who didn't understand anything that was happening, and kept repeating that she was going home to Glasgow. Social Work had managed to find her a week's emergency respite at a care home nearby, and were packing a small bag for her. They would try and find something more permanent as a priority. At this point, John, who had been listening, spoke.

'She can't come back here. We've done our bit when no-one else would. This is killing my wife. We can't work, we can't socialise, and it's hard to pay our bills. Lorraine's on steroids to keep her heartbeat regular, the kids are frightened and feel neglected. This is stopping today.'

He took my hand in his and looked me in the eye. 'None of us can take any more.'

As the tears streamed down my face, the social workers quietly told Mum she was going on a little holiday, as my health was bad and I needed a break. They were taking her away.

Suddenly she reached out and slapped my face.

'How dare you do this! I hope *your* daughter does this to you!'

She cursed and cried as they took her arms and marched her out to the waiting car. They didn't have to put her in a straightjacket, but apparently they had one in the car in case, they told me later that week. In the back seat, she yelled at the window and tried to open the rear door as they drove out the drive.

Writing this now, some ten years later, I still find it difficult to put down how upset, exhausted and utterly defeated I felt at that moment.

7.
Beginnings and Endings

My friend Jacqui, who owned a lovely farm and luxury B&B on the south coast, had previously helped John and I erect our polytunnel. She came over to mop up the tears and help me try and make sense of it all. Jacqui is a breath of fresh air; perhaps not the most business-like of people, but if you are in a spot, she is a great lady and a good friend. Sometimes, she even comes out with really useful ideas, although she may be totally unaware of doing so at the time.

And so, as a result of a random conversation over tea and biscuits, I decided to embark on a fact-finding mission to find out about possible "exotic crops" to grow in our greenhouse. We needed to try and make it more productive, perhaps with some sort of cash crop to bring in a little more money. John had mentioned a character, who worked as a welder beside him in Newquay, who grew chillies and the like for sale to the Indian restaurants in Newquay. We went to see his set-up and it became immediately clear that Eddie (not his real name) was growing a lot more than just chillies.

His polytunnel was stuffed with rows and rows of cannabis plants in grow-bags on the floor. There were, I have to add, a good amount of potted chilli plants also, but the fact that the cannabis plants were in different stages of growth, told me that this was his main "crop" and obviously he had a thriving business. Newquay was his main market, with many stag and hen weekends. He had a constant demand which he couldn't really supply, and was keen to have another grower on board. A few seeds were thrust into my hand and he told me propagation was simple, and to get in touch, when they were budding. He'd take care of the rest he said, tapping his nose with his finger. 'Keep mum', he said.

John was so reluctant to have anything to do with this that I passed the seeds on to an acquaintance living on the outskirts of the village. When I dropped her daughter off from school some weeks later, she proudly took me to see her polytunnel. Inside cannabis plants were almost touching the roof.

'Are they supposed to be that tall? Eddie's weren't that tall.'

'They just keep growing,' she said. 'Healthy enough, but no sign of buds or flowers.'

'Have you… tried any?' I asked, noting that there was no smell coming from them.

'Nah, waitin'. Can you ask him if this is how they're supposed to be? Take a photo.'

So, I grabbed my mobile phone, took a photo and sent it to Eddie. My phone rang almost immediately.

'What's she bleddy feeding them plants with? They look about 6 feet tall!'

'They are! They're touching the roof of the polytunnel.'

'Ach, Christ, they must be *males*!' he spat.

'So, what does she do now?'

'Not worth anything really, unless she wants it for hersel',' he said trying and failing to keep the disappointment out of his voice.

'So, she'd be as well cutting it all down?' I said, watching her vigorously shaking her head.

'Try again next year. It's the germination; sometimes if the temperature isn't right, you get all male plants.'

I went back to see her a few days later and she was indeed cutting all the plants down and putting all the greenery into black bin bags.

'Bleddy daft idea!' she said, 'I'll get Merv to put it all in the incinerator with the rest of the garden waste.'

'Er…you're going to burn it? In your garden? Don't you think it'll smell?'

'But it's not the right stuff! It'll be fine.'

*****

I received a phone call two days later when she could hardly talk for laughing, telling me about how Merv had phoned her whilst she was working at the supermarket at the other end of Wadebridge, to look over to the village, and all she could see was a huge plume of grey smoke clearly coming from his incinerator in the garden. He had doused the fire after a few minutes, petrified the fire brigade would turn up and the whole business exposed to the world.

'I had to take the other three bin bags-full to the tip. The nice man there was going to help me, but I said, never mind, the dog had poohed on the carpet, and this was me cleaning up!'

Village gossip being what it is, I heard a few months later that she had tracked Eddie down herself and grown on some cuttings which did so well she was supplying not only the younger aficionados in the village, but had a steady stream of customers coming from Bodmin and was making enough to give up her supermarket job. Our own polytunnel was strictly confined to growing tomatoes and salad. The lucrative cannabis market was one we just avoided after all the excitement and at our smallholding, we concentrated on trying to go back to what we knew worked, what we had done prior to Mum moving in.

Dave, our friend from the village, had been appearing

more frequently as he was a bit lonely following the death of his brother, and as I wanted to start a proper vegetable garden, I offered him paid work to clear it and plant it up for us. John offered to rotivate it all, as Dave was getting on a bit, and Dave was delighted. So, once a week he would appear, do a bit of work, and we would treat him to cooked dinner or rolls and bacon or whatever. He was happy he was making a bit of cash and didn't feel as lonely. The freezer filled with French and broad beans, jars filled with beetroot and pickled onions, and strings of big, fat onions would be hung up to dry. It was a great arrangement and suited us all. He was a tower of strength when I would tearfully return from visiting Mum in her now permanent care home and give me a rough hug, and tell me I was 'a fine dattur and no mistake.'

We tried to reconnect with the children, but the dynamics of the relationship had changed forever, and they were closer to each other than to us. They were growing up and the combination of work, Mum's illness and teenage hormones, all conspired to keep us from regaining the balance we had before. Even though my mother was no longer living with us, her memory hung around the house like damp, and the duty visits I made to her nursing home conspired to keep me depressed and edgy, whilst the smallholding fought back to keep me sane and happy.

My favourite place was the orchard. It seemed to have a timeless peace and a few hours working there restored my calm. In turn, the orchard rewarded me with more and more apples and, as autumn approached that year, we planned to increase our Spotty Dog Cider production. Some locals with orchards offered us their apples in return for a small amount of juice or cider, which we readily agreed to. Volunteers helped us to harvest and start the pressing for the new season and the familiar sickly-sweet smell of fermenting juice filled the air.

Meantime, my brother and I were forced to sell Mum's house to pay for her care, as the fees had swallowed up all her savings. Poor Mum would have been devastated to know that everything she and Dad had worked so hard for was fast disappearing in care home fees, but thankfully she was so consumed by the disease at this point that she was living in a half world of memories, reliving her childhood.

And so as the saying goes, life goes on and the wheel turns. Everything on our smallholding was cyclical and we tried to work with the seasons. After February lambing time, spring arrives quickly in Cornwall. The year starts with the sloes blossoming along the hedge next to the lane, covering black twigs with snowy petals. Before you know it, pear trees are joining in to swell the spring song with lemony-scented blossom, and a gathering crescendo of honeybees fills the tree with humming, droning and movement. Winds are still brisk and sharp, sending flurries of white petals into the air, but the days start to lengthen and gradually winter pulls his frosty cloak around him and flounces back northwards. Spring shakes out her green gown sprinkled with dandelions and daisies, and settles herself to linger a few months.

To cheer myself up I offered my services for lambing help at the farm in Launceston for the following spring, and asked to have a couple of lambs to take home as payment. The brothers who ran the farm were pleased, as even with my little knowledge, every pair of hands was useful and I would be used to bridge the time between shifts to allow them to go home and have a bath, eat and grab a short sleep. Lambing time is extremely stressful with long physical shifts. Add into the mix capricious bad weather or illness in the flock, and farmers can see their profits disappearing in front of their eyes.

The plan was to keep a couple of lambs from March to October when they would go for slaughter and fill our freezer. We wouldn't have any shearing, winter feeding or breeding issues, but could have the lambs young to bottle feed and cuddle. They would keep the grass in the paddock down all summer, with a little added protection for each apple tree. Financially, it was a great idea too. The cost of the replacement milk feed and the cost to slaughter and butcher was less than a third of what we could sell a butchered lamb for. So one carcass would go into our freezer and the other sold to friends. And so we managed a really sustainable option that suited everyone and I had my lambs to cuddle.

As the days warmed up and lengthened, pink and white apple blossom popped open over a few months. Some early varieties of apple, such as Discovery and Katy, finished and set tiny fruitlets before the late varieties had even opened. The Miscanthus greened up again, quickly thrusting sharp spikes high into the air and deadening the noise from the A39 half a mile away. Easter saw the return of tourists and holiday home owners, and the sales of Spotty Dog Cider rocket. More pubs were keen to stock it and we watched the bank account swell.

For my birthday John made me the most amazing and beautiful weathervane for the top of the shed. He spent a lot of time in the forge, fabricating all the pieces required, and drawing a cockerel on sheet steel and cutting this out. With black direction markers and arrow and a golden cockerel on top, this marvellous gift was almost ready to be fixed on the top of the shed, so we could see it from the kitchen window. I was amused to see him holding it up and blowing as hard as he could at the cockerel.

'What are you doing?' I called.

'Making sure it's moving freely in the wind,' came the answer. Of course. How could I not see that?

It was a magnificent sight, shining in the sun when he finished erecting it the next morning, and was a useful indicator for us on the top of that hill at the mercy of cold northerly winds or wild South-Westerlies bringing gales and rain.

Customers phoned to book their garden machinery in for servicing in ever growing numbers, meaning John was busy every day from 8am till 7pm. He only had Sunday off because I insisted we have some time together as a family; although the children now had their own lives and were off doing rugby or swimming in the sea with their friends. Neither of them were interested in helping out on the smallholding, and our thoughts inevitably turned to the future and when they'd fly the nest.

As summer days lengthened, our days grew longer, with us both working at 7am and not stopping some days until after 9pm. Anyone considering life on a smallholding needs to understand that you get jobs done when the weather and light allows, and according to need. Summer is an incredibly tiring time as the days are just so long and the increase of livestock numbers and tasks due to births or hatchings and subsequent increase in cleaning housing adds to normal day to day chores. Animal welfare comes first, and this doesn't only mean turning animals out and feeding, but checking them for health and ill health, vaccinations, mending fencing to keep them secure, and in high summer, ensuring they have shade and adequate clean water. A single chicken can drink a litre of water daily in hot weather and drink dispensers frequently film over with green algae. We recycled empty plastic milk containers because they were easy to refill and when they went disgustingly green we simply replaced.

Kate decided she was going to do the Padstow to Rock charity swim in July. This charitable event to raise money is an annual Cornish tradition. This was her first attempt and she was practicing frequently. She was a strong and confident swimmer and on the big day she donned her wetsuit and took just under an hour to complete the one mile crossing. I'm proud that even though my kids perhaps didn't have everything they wanted growing up, they appreciate that they have more than some others.

We sat outside that evening with some cider, watching our bats emerge from the roof tiles. They were doing a great job eating insects as they flew and by switching on the lights near our little lily pond they would happily fly very close to us, collecting the moths attracted to the light. Our neighbouring barn owl took his evening flight silently through the orchard and perched on the feeding post John had erected for him along the fence of the orchard. Here, we had left a strip of rough grassland to let him hunt for voles, his favourite food. He had plenty of great hunting areas across the lane at the iron-age hill-fort, but we liked to see him in the garden. We thought we had seen two different owls hunting earlier in spring as the flights had been frequent, and assumed he was feeding youngsters, but we never saw them.

After midsummer day, the bees, clever creatures, always recognise the change in the daylight hours and whilst the queens still lay a lot of eggs, the workers start to build up their store of honey. Tiny hexagonal cells are cleaned and polished and filled with honey, before being sealed under a little wax lid, and the frames in the honey supers are soon groaning with weight, smelling delicious. A heady perfume of wax, heavy spicy clover and light, aromatic fruit blossom surrounds me as I work the bees.

As the days shorten, the drones, the male bees, serve no further purpose, and the workers remove and exclude them from the hive. With the prospect of a long, damp winter, extra mouths to feed are the first casualties from the colony. We took the hint, and removed a small amount of honey for bottling for ourselves and some cut-comb honey to pack and sell. The hives are then given a varroa treatment to ensure as many parasitic varroa mites in the colony are killed and don't weaken the bees. These tiny mites are now found in bee colonies all over the UK and good beekeepers try to prevent the population from becoming a burden on the bee colony, which weakens them and leads to all sorts of problems. We leave the bees with a good amount of stores to see them through the winter until March arrives with warmer temperatures.

Autumn is a time of consolidation and harvest and our attention turned to the sheep, who were tempted and encouraged into the trailer and taken the short journey to the family abattoir a few miles away. We would return in a week to collect the meat, ready to freeze, produced from free-range, grass-fed animals who lived happy lives with no stress. Although some people may have issues with producing meat, these animals were well treated, and contributed much to our sustainable lifestyle. They kept the grass down and we didn't have to use petrol-powered grass cutters, so they were also reducing carbon.

The Rural Business School courses were going well, with a big uptake, and new learners embarking on courses. I put together even more diverse courses and training, including a Beginners Horse Logging course, taught by a wonderful couple from West Cornwall and their equally wonderful and intelligent logging horse. The feedback was fantastic and another course quickly arranged. However, all was not rosy on campus.

Rumours flew around the college that EU funding wasn't being renewed on the same scale, and there was uncertainty over funding and possible restructuring. Meanwhile, the Rural Business School was awarded The Queen's Anniversary Prize for Higher and Further Education; a fantastic achievement and boost for all the staff. The staff came from all walks of rural life: from farm families and

backgrounds, agricultural universities and some, like me, from humble smallholdings. The varied roles of the staff meant we could offer training, business support and applied research all over the land-based industries from the aspiring smallholder to the large food producing industries in the South West. You can find out more by looking up
www.ruralbusinessschool.org.uk

John was delivering a variety of courses, but the demand for blacksmithing was highest, and it was a joy to see him happily passing on his skills and experience to many, from youngsters trying to get a start in the trade to fathers and sons who did the courses together as a birthday or Father's Day treat. His father had been a time-served blacksmith and was the last blacksmith to shoe a working horse in our home town of Paisley, and it felt right to see the son following the father in the magical art of forging and metalwork.

It was lovely to watch the acrid smoke from the newly-lit forge drifting out the forge door on those mornings. It would take about an hour to get the forge hot, and ready to work and students would arrive in time for a coffee to accompany their safety briefing. Then the hammering would last till lunchtime, when most students were delighted to rest their hammer hand and be ravenously ready for lunch. A hot Cornish pasty and cool drink satisfied them, and they were ready to continue afterwards with their projects.

By the end of the afternoon the students would proudly display what they had made; normally a fire poker, a hook and a doorknocker. We even had a booking for ladies from a local W.I (Women's Institute) group. We were a little apprehensive as the youngest member was in their late forties and the oldest in their eighties, but they had a great time and were very competitive. We took lots of photographs of them during the process so they could have a presentation evening later in the year, show off what they had made and how they made them. Usually a quiet man, John really opened up during those days, which always ended with an evening trip for us both to the pub, which had been transformed from a dirty, empty, unwelcoming place to a vibrant and busy meeting place in the village.

The new owners had brought the pub back into the centre of the community, which was a great thing for a small village. New indoor toilets had been installed, although some older locals bemoaned the loss of the old character of the place. Kieran got a part time job, first in the kitchens and then behind the bar; and for a time, our very own Spotty Dog Cider was also on tap, which was lovely to see and very popular with the summer tourist who flocked to the renovated bar and restaurant.

Cider bottling was stepped up with a few close friends helping us to bottle, pasteurise and label our bottles. Labelling was only done in small batches through the winter months, mainly because the shed was freezing cold, even with the wood burner on all day, and our fingers and feet became numb very quickly. We kept the cider in distinct batches, as our customer feedback revealed that some like dry, still cider and some wanted it sweeter with a bit of sparkle. I took up offers of help and instruction from many kind producers in the midlands and created my own specialities. Mind you, it's not an exact science and is dependent on accurate measurements of the sugar levels at the start and end of the process, of making sure your cider has fermented to dryness if this is what you need, and blending and maturation at the end to result in, hopefully a delicious product.

At the end of a very long and tiring day, I brought some bottles in for my trusty testers to try. John has a bit of a sweet tooth and particularly liked the Quencher batch, a medium still cider made with mainly dessert fruit, and around 6% alcohol by volume. My son, however, declared that to reach the younger market it needed fizz.

'Ta-dah!' I declared, and pulled a bottle from behind my back. 'I took your comments about this on board and have prepared a test batch of a few bottles. '*This*,' I declared, 'is Quencher with fizz!'

I theatrically released the cap from the bottle, only for the cider to shoot to the ceiling with incredible force and showered the wood burner, myself and the walls and floor with golden rain.

They both stared at me and then started laughing hysterically!

'That wasn't supposed to happen!' I said in horror, looking at the remaining half bottle still fizzing hugely.

'I'll get you the mop then,' said John, leaving the room, which was starting to smell very appley. 'You'll have to wait till tomorrow to clean the wood burner when it's cool'.

'Shit! I'll need to adjust all the measurements for charging now!' I said and headed for the cider shed to writ-up the failure. Thankfully, I'd only made ten experimental bottles, and easily reduced the charge for the next batch, which I'd test by opening outside in future.

Cider-making is all about trial and error (and keeping notes). I managed to create the perfect recipe after the fourth test batch. Kieran told me it tasted exactly like a well-known commercial cider. Praise indeed.

The Wassail we had planned for mid January had to be cancelled, due to the heavy rain making the orchard unsafe for a crowd of people, and the weather continued miserable. We had been looking forward to a small bonfire and to toasting the orchard with our own cider, but it wasn't to be. Instead, John and I spent a quick twenty minutes in the orchard pouring a libation, but retreated to the house and the wood burner to drink our cider.

Sales started in earnest at Easter. We had four local pubs

regularly taking cider to sell at the pump, and lots of bottles to sell at Padstow Food Fair, a small artisan food market held a few days a week in a very popular tourist town on the North Cornwall coast. Feedback was great, with lots of returning customers and prospective makers coming to talk to me at the market about how they made their cider.

Whilst the cider was doing well, the bees sadly, had not all made it through the damp winter and by the middle of March, it was clear from the lack of activity and flying bees, that we had lost the weakest colony of bees, as I had feared. When I opened the hive a very small cluster had clumped together at the bottom of the frames in the rear of the brood box. A combination of lack of food and heat had finished them. It was a horrible chore to remove all the dead bees and the wax foundation in the frames, but beekeeping is all about good hive management, and we needed to replace the foundation in the frames in the brood box to prevent anything affecting a future colony. The remaining colonies were strong and busy, with workers darting in carrying little bags of bright orange pollen on their legs from the earliest dandelions and crocuses. The collection of pollen is a sure sign that the queen was alive and either laying or getting ready to lay, and we hoped, if the colony swelled enough we might be able to create a new colony for the now-empty third hive.

At the Rural Business School things were changing, as we had feared, and there was a drop in European Union funding. Staff had not been replaced as they left and there was talk of redundancies. I decided to resign to concentrate on my cider making. After all, I'd had a wonderful three years there and wanted to leave on a positive note.

At a marketing event, it was suggested that I enter the Cornwall Sustainability Awards, and I had a big discussion with friends about what this entailed and could mean.

Although we had always lived on the smallholding as sustainably as we could, this really stemmed from lack of money, and a reliance on the inter-connectedness of what we did. It just seemed like common sense to us to live this way; and I felt a bit embarrassed to apply for an award where we were just doing what had evolved naturally. However, my business sense said it would be a good process to go through on paper, and might even highlight things we could improve on, so I completed the form, wrote a brief paragraph about what we did and why, and emailed it off.

We managed to make time to take long walks with Lily, our Cocker Spaniel, along the beach at Rock and also to Port Quin, where we climbed the hill, joined the Coastal Path and sat watching dolphins and the most wonderful sunsets. Those few days we managed to sneak away from the drudge and hard work allowed us to sit and talk and to rediscover what we liked about each other. One day we coerced the kids into coming with us, packed a picnic, and put two canoes on the car roof-rack. It was great to hear them shrieking as they canoed out to sea, or when they decided to fall in and swim amongst the sea bass in that crystal clear water. We ended the day with a picnic on the rocks and when the sun started to dip, headed back home. It had been a magical day.

Autumn finally arrived with wet, blustery winds and Kate going off to University for the first time. She had gained wonderful grades at college and was ready for a new adventure in a new city, and I envied her starting on this new phase of life. She needed the stimulation of a different, vibrant place to live and to make new friendships, and I wondered if she would ever come back to Cornwall after having enjoyed the delights of a bigger world. I was under no illusion that this remarkably clever and pretty young lady would soon make her mark on the world.

We didn't have the easiest of relationships, both of us being strong characters, and the past years had not been easy with all the work, worry, and Mum coming to live with us. With the benefit of hindsight, we both may have made allowances for past behaviours, but we made the best choices at the time and I hoped that in future we would become closer. After all, I was living my life the way I wanted and respected the fact that her life in turn would be her choice and her way. And so, yet again, family dynamics changed and we all changed slightly as a result.

I started to consider our lifestyle and whether it was time to call it a day. Whilst we really were living the "Good Life", it wasn't all idyllic. There was the exhilaration of working outdoors in the sun, collecting and transporting apples to the production shed to be pulped and turned into golden, heady juice and this felt so right, but was very hard work. That particular golden summer was perfect; with showers of gentle rain to plump up the apples, followed by warm sunshine, and we enjoyed the occasional evening trip to the inlets at Port Gaverne to fish for mackerel.

With the end of the summer there was no rest, as we started harvesting the apples in the first week of September. The sunny summer ensured a high sugar content and the cider we would produce looked to be around 6 or 7 % alcoholic strength. We collected some old Cornish apples from an orchard at a nearby manor house and pressed them as a special batch, which would be called *Reserve*. It would be matured longer than our normal cider, and promised to be delightfully floral with a light first taste, deepening to an oaky and vanilla finish.

This all sounds very like wine tasting with pretentious words such as terroir and malolactic fermentation, and taste notes such as straw, grass, butter, spice all being bandied about; but there really is just as much variation in cider-making as there is in wine production. Naturally, different varieties have their own characteristics, and whether the year has been sunny and dry, or cold and wet, affect the taste too. The blending, if there is any, at the end of the process, and how long the cider is allowed to mature also affect the taste.

When we attended the local Cornish Cider Festival in Lostwithiel that autumn, the organisers asked me to join their taste panel for the amateur cider-maker competition. Well, I have to say the standard was very high and there were a lot of entries. One of the ciders that year was particularly delicious, and indeed won the competition. It turned out this was the winner's second year as winner, and we had a long chat about what he would produce in the future. It was nice to talk to fellow enthusiasts; after all, the only difference really between me and them was that I was making a higher volume and selling it.

Our lambs that spring had developed into fat, healthy sheep and would soon be gone, leaving the paddock to the chickens, the bats and the owls. We sat in the small enclosed garden at the rear of the house next to the lily pond that Kieran had created for us, drinking wine and cider, watching the Milky Way above us and the ghostly barn owl as she hunted for moths amongst the apple trees laden with a superb harvest. We talked quietly about everything and nothing. Thankful for the blessings we had and content that for the time being everything seemed to be going well. Of course, being at the top of the wheel means that, inevitably, the wheel will turn. Autumn is a funny time of year and whilst it is a time for being thankful, it's also a time for reflection.

Mum had deteriorated physically and was bedridden, following a serious fall, and was now being made 'comfortable' in her nursing home. Her dementia medication had been withdrawn as the consultant felt it was no longer having any effect, and I braced myself for the inevitable as she entered the final stage of dementia. I still visited twice a week, although she was often sleeping, or failed to even notice I was present. The care home staff told me not to bother coming, but I felt guilty if I didn't visit and depressed when I did. Kieran felt uncomfortable visiting any longer, wanting to preserve the fond memories he had of his gran and so, occasionally John would come with me for a little emotional support.

It's heartbreaking to watch the disappearance of a loved one. The shell is still there, older, feeble and reminiscent of the person they once were, but when they open their eyes, there is no recognition, no awareness, no essence of the personality left, and yet the body still clings on; like an old machine, gasping and stuttering in the last cycles of its use. My visits were mainly to check she was being kept clean, comfortable and not left alone. The staff and I discussed and agreed an End of Life Plan, and they told me it was simply a waiting game.

On the smallholding, we were working ridiculously long hours, with John pressing and juicing, and me harvesting and transporting apples in worsening weather. I was starting to feel very tired and old, and wondered if we should be taking it slightly easier. This year we had made 4000 litres of cider and, frankly, it was physically too much, even with two volunteer helpers. We should have employed a couple of seasonal staff, but we couldn't afford it, and making such large quantities was worrying. Thankfully, we only lost one 200 litre batch which was tainted and smelt and tasted of pear drops. As I poured it away, John was almost crying at the waste and the lost revenue. But, the cider had been increasingly popular, with huge sales starting just before Christmas.

In early February and after a long battle with dementia, Mum finally died. She fought it to the very end and we were frequent visitors to her bedside. There was of course, nothing we could do, except sit and gently talk to her, reassuring her she wasn't alone.

They say that when you have a loved one with dementia you have the grief of losing them twice; once when they cease to be the person they once were and again when they die. It's a tortuous journey not just for the person with the disease, but for all those they leave behind. When the dreaded final phone call came, John and I rushed to ensure we were there with her at the end; and it was with a mix of emotions that I sat holding her hand as she quietly and finally relaxed, at last at peace.

8.
Decisions, decisions, decisions

Lambing time in mid February provided me with a much needed distraction and sense of purpose. I delightedly cleaned, disinfected and bedded up new pens ready for the new arrivals. Our farmer friends were glad to have my help and we slipped into a regular routine as we had done for the previous lambing sessions. I would turn up at teatime and take over from Rodney, who would pass on anything I needed to know, including which lambs needed feeding, where they were, and who to look out for. As they were lambing 600 ewes in three different barns I was never at a loose end, but the return was great therapy. At midnight, I'd hand over to Julian with the list of new arrivals etc. Every year I managed to collect some injury and this year my right shoulder had been wrenched trying to catch and get down a particularly belligerent North Country mule. It was her first time lambing and she really didn't want any help, but with just a head showing, needed intervention. Finally, after a good fifteen minutes chasing her round a confined pen, I managed to grab a good hold of her, but almost dislocated my right shoulder in the process. My annual visit to the osteopath eased the damage, but I had to start thinking about lambing as a young person's occupation, and I decided at the end of lambing, this would be my last season.

Dave, our friend and vegetable gardener had been

hospitalised with a small stroke and we visited him frequently to cheer him up, explaining that we still expected him to look after the garden when he recovered, but that I would have to do the gardening, with his supervision. He was glad to still have a purpose and his recovery was quick, and soon he was back home, almost back to his usual self. Our other friend, Colin made sure he collected Dave and they went to farm auctions as before, albeit for shorter visits to stop Dave from becoming too tired. This is the nice thing about small, rural communities. Everyone knows everyone's business, and the majority of people help each other in times of need.

I found coming to terms with Mum's death very difficult. Despite the fact that we were not particularly close, losing your mother leaves a hole in your life, and no matter what I tried to plug that hole with, I realised only time would heal it. Anyone who has been bereaved will know there are many stages you have to work through and they don't always appear logical at the time. This affects everyone around you also, and is hard if they too are working through their own stages of grief. Of course, for the living, life has to go on; animals have to be fed, medicated and sold, and money has to be found to pay bills. Auto-pilot offered a refuge for many things but not all, and I found it impossible to try and "go back" to how things were before Mum had come to live with us. The whole experience had changed us all. I couldn't face going through her personal things just yet, but John and I removed all her furniture from our annexe, where she had lived before moving to the care home. Once cleared and redecorated, we successfully advertised the annexe as a short term let, which provided a good income and allowed us to have a short break in France in spring, our first holiday in eight years.

In March, dandelions popped up everywhere on our lawn, so I thought I'd try my hand at making dandelion beer. If you want to try it, the recipe is here. Dandelion beer was a new and successful experiment, but Kieran had watched me preparing this and would not try even a drop of it, after watching me boil whole dandelion plants in my big jam pan. Okay, it looked a little bit green and unappetising in the pot, but undaunted, I bottled it and it turned out fine, if a little 'boisterous'. Keeping it in the fridge controls the fizz slightly and it's best enjoyed cold as a refreshing antidote at the end of a long day.

****

Dandelion Beer

8 oz whole flowering dandelion plants with big fat roots
8 pints water
1 lb sugar
¼ -1/2 oz root ginger, peeled & well bashed
3 or 4 raisins
1tsp cream of tartar/tartaric acid
1 sachet white wine yeast
Rind & juice of 1 lemon

Dig up whole young, flowering dandelion plants and wash well. Place in a large jam pan with the ginger, lemon rind & water. Boil and simmer for ten minutes. Pour into a fermentation bucket and cool till warm. Add the sugar, lemon juice & tartaric acid & stir. Add the yeast, cover, and leave in a warm place for 4 days. Strain off the liquid and leave to settle. Then siphon into clean, screw top bottles. Leave in a cool place and release the pressure FREQUENTLY (at least twice a day). Ready to drink in 10 days, and best within the month.

****

A few days later we had unwelcome visitors. John was in the new shed working on rebuilding an outboard boat engine on the bench when Lily, our Cocker spaniel started barking and growling. I looked out of the kitchen window to see a worse-for-wear white transit van parked inside the gate, and two characters walking around, looking at the sheds.

I walked out to confront them, closing the bottom stable door as usual. Lily was inside and barking constantly. This seemed to deter the two men, the older of whom approached me slowly.

'Just looking to see if you have any scrap missus,' drawled an Irish accent.

I was instantly wary, identifying them as tinkers or travelling folk. There had been rumours of their arrival in the area in the last week.

'No, we don't have any scrap,' I said, but still they lingered.

'Could have a look in your shed?' he said smiling, and walking towards the shed, but I was already there, closing the door to prevent him seeing our expensive machinery.

'There's nothing here for you,' I said firmly.

Still they stood looking about, eyes everywhere. I was aware that John was inside and probably could hear the exchange through the wall, and so I felt safe. Ish. The second man spotted some old tractor batteries lying in a heap outside the shed and headed towards them.

John miraculously appeared.

'Nothing here for you friend. On your way or I let the dog loose.' He stood calmly wiping his hands on an oily rag. He's a big man and easily dangerous when threatened, but calmly and quietly stood and faced them down. The two men headed slowly back to their van, eyes still assessing everything, possibly looking to see what they could come back to take at night. You could have cut the atmosphere with a knife.

'Lil!' John called sharply, and like a rocket, the snarling, barking black and white tornado of a Cocker Spaniel leaped over the shut stable door and flew to John, standing square in front of him with teeth bared and hackles up. All thirteen inches of her. She might be small, but by heaven, she is very territorial is Lily. The message seemed to finally have been received and the men quickly jumped in the van and drove off.

To ensure no unwelcome returns, we padlocking the gate at night and any time we were out. You can't keep thieves out, John reasoned, but you can make it damn difficult for them, and sadly, farms and smallholdings are easy pickings for the theft and resale of farm equipment. As John's main business was servicing garden machinery, we were aware that it might not just be *our* machinery and equipment being stolen! This was a bit of a wake-up call for us, and soon after, we installed CCTV and also improved our home security.

Holiday time arrived and we had our bags in the car, about to drive to Bristol airport for our few days away. I handed Kieran a sheet of notes and phone numbers and a wad of money, warned him not to have any parties when we were away and to remember to lock up when he left the house. Secretly, I'd asked a couple of friends to look in on him but was sure he would manage, as at heart, he was a sensible lad.

Our break was wonderful. We arrived in Bergerac in the South West of France and hired a car. John did all the driving so I could just sit back and enjoy the scenery. It was such a vast and beautiful country! The weather was sunny, but not overly warm, and we walked, and talked and began to relax. When we returned to rainy Cornwall a week later, Kieran asked,

'Well, what did you do?'

'Nothing really. Just relaxed.'

'What did you see?'

'The countryside, and some chateaux and villages. Nothing much.'

My son looked surprised and a little sad for me, I thought.

'I don't know why you bothered going Mum. Was it *so* bad?'

'No Kieran, - it was wonderful,' I said smiling.

'No, it really was nice,' John answered smiling too.

Kieran looked at us both and hugged us.

'Well, I'm glad your home,' he said. '*And* I didn't have a party.'

All looked well as we walked round our small acre that evening. The hens were happy, the orchard looked good and was starting to show a touch of green, the sloe in the hedge was blossoming and filling the air with its heady and soporific scent, and we prepared to get back into the saddle again the next day. We felt rejuvenated after our short break and looked forward to the awakening of the countryside. Change was in the air, as if the returning spring had decided to sweep away the darkness and gloom.

Spring turned to summer and the county blossomed with hoards of holiday makers arriving, visiting the beaches, clogging up the narrow lanes, and emptying the supermarket shelves of essentials. The land finally dried up and the fine weather seemed to suggest we were in for a fantastic summer. The *Doc Martin* film crew and actors arrived in Port Isaac ready to start filming another series and the businesses in the village were thriving. I was selling lots of cider, delighted to see my bank account looking healthy. The latest hatching from the incubator were finally ready to be released outside into their juvenile pens, the hens and cockerels moulted their old tatty feathers and the paddocks were full of small feathers dancing in the air. A late evening stroll to shut the gate revealed two romantic hedgehogs running round the grass together, creating quite a disturbance for their size. I suppose when two excited spiky pin-cushions decide to "get it together" the margin for error is pretty small and this may explain the frequent unromantic squeaks, growls and grunts. Country living was glorious, I reflected with a smile as I returned to house.

Part of country life was the tradition of the hunt. The refurbished pub was hosting the North Cornwall Hunt meet that weekend, and we combined having a drink with watching the huntsmen in their smart red and black jackets and hounds milling around the pub car park. No matter what the letter of the law said, it was obvious that local customs still went on in Cornwall, and that the local gentry, farmers and country folk alike were cheered by the sight of huntsmen enjoying their stirrup cup with hounds milling about.

A few days later, whilst I was weeding my vegetable garden, I could hear the baying of the hounds coming nearer and nearer. They sounded very close, so I walked round to the lane and saw the hounds in the unfenced Miscanthus, a mere hundred yards away. Not good. I hurried round to the feed shed and grabbed a bucket of chicken feed, knowing that if they got in over the fences we would have a problem with far too many silly chickens in the paddock being fair game for dogs in a pack.

There is nothing as stupid as a chicken, well, possibly a duck, and no matter how I tried I could not get the hens to go into their houses as quickly as I would like, despite chucking handfuls of corn inside and rounding them up as best I could. Finally, most of the birds were in, and the ducks too. As I shut the door of the duck house the first hounds were in the paddock, and I ran, flailing my arms and shouting at them to sod off. The lambs were frightened and bleating and thankfully headed into their shed without any problem. A huntsman was on the other side of our boundary wall with his horn, trying unsuccessfully to bring the dogs back under control.

Angry words were exchanged and, finally, he managed to get them off our land. I ran round to the front of the house and onto the lane, where my neighbour was shouting at the Master in very colourful language that they were trespassing and that she would report them (I'm tactfully omitting the volley of swear words she actually used). They eventually turned down the lane to try and prevent their wayward pack from wandering onto the very busy A39. Only when the barking was very faint did I feel it safe to release the birds and the lambs back out. So, another busy day had passed with nothing done. Weeding would have to wait till another day.

We managed to go out more frequently, and in late spring increased our fishing expeditions to nearby Port Gaverne, where we caught mackerel for the freezer and mingled with the film stars. The actors from the TV series *Doc Martin* often had a drink or two (or three) in the Port Gaverne Inn after filming. The pub was very quiet off season, but in July and August we avoided it, as the lanes were full of holidaymakers who couldn't negotiate narrow lanes, and filled the car parks. The trick during "silly season" was to go out early and do whatever you needed to do and then return to the smallholding.

Jaimie, our ginger cat continued to play tricks on us with the little 'gifts' he brought us from his hunting trips around the smallholding. We received packages on the doorstep daily. From this we deduced that there was a healthy vole population, which was borne out by the equally frequent sightings of our local barn owl; but he occasionally brought us surprises, such as the odd mole and even a weasel, which must have been a bit of a challenge for him. Thankfully he never brought us slow worms or snakes, which we knew were common in our area. When we took the dog for a walk along the lane and into the remains of the Iron Age hill fort, he would accompany us, following behind and trying to keep up with his little legs. If the pasture was too high, he would sit on a fence post and wait and watch for our return.

The insistent meowing we often heard at the front door heralded the arrival of a token of affection from him, although we became wary of allowing him entry until we have done a visual "mouth examination". Late one night when we had just gone to bed, John was woken by his meowing and grumpily went to let him in as usual. Jaimie sauntered in, brushing past John's naked legs, and opened his mouth, depositing a very live mouse onto the hall floor. The mouse started to run around John's feet, looking for a place to hide, causing a minor panic. He let out an exclamation and picked up a nearby Wellington boot and repeatedly tried to hit the creature, whilst Jaimie looked with apparent amusement. Hearing the rumpus, I also got out of bed to view the unforgettable sight of a slightly chubby, middle-aged man in his underpants, dancing around the wooden floor, swearing and contorting as he flailed around with a big Wellington! Calling the cat the most useless mouser in the world seems to have done the trick, and finally Jaimie pounced, and when the mouse was firmly inside his mouth, albeit with a tail protruding, John picked him up and chucked him outside. There is *never* a dull moment in this house.

As local beekeepers we are always looking out for swarms and had a phone call asking us to collect a swarm of bees nearby. So we filled the car with suits, some bits and bobs and a cardboard box. A few minutes later we were looking at the small swarm on the side of the chimney of an old cottage next to the road that went up to the village. The owners had a tall ladder, but it wasn't the easiest swarm to catch. John ascended the ladder and brushed all the bees into the cardboard box. A few cars had stopped to watch what was going on by this time. The problem was how to come down the ladder with the box. He managed with difficulty to get about half way down, to about twelve feet from the bottom, and then turned, with his heels on the rungs. This wasn't the best idea and soon he was half sliding and half running towards the bottom. The owner and I realised this could result in the bees being flung out, so we retreated to the other side of the road. But miraculously, John landed with a spring, onto his feet, and the box intact with bees. A round of spontaneous applause from the cars, and he took a small bow before we got into the car with our prize and headed home.

The bees that year rewarded us with over 60kg of beautiful golden honey, rich with the heady scent of summer. This sold well to our regular customers, who loved eating a natural product, with the only air miles involved in its production being the ones the bees themselves flew. The weather held and the bees started to make honey from the blackberries, the honeysuckle and the flowers of the ivy. We would leave this with the bees to help them over the winter.

With the arrival of autumn we prepared the cider equipment and again discussed the possible expansion of the cider business, but John really couldn't spare the time away from his own busy machinery business and we were both just so tired. To expand meant we would have to employ staff and upscale our equipment, and I was very wary of this. I loved what we were doing and maintaining the personal involvement of it all. At my age, I didn't want it to turn into a pressure driven big business.

At our current size, I could tell people exactly what the varieties of apple were in their bottle of cider, where they came from and explain about our practices on the farm. How the orchard was shared with the hens who ate the bugs around the apple trees, about the barn owl who took the moths at night, and about the wildflowers allowed to grow there without any need for fertilizer or pesticides. This seemed to be, in addition to its delicious taste, what people wanted to buy. They were yearning for that rural idyll, and we could sell it to them in their cider. Our website photographs showed the animals and birds in the orchard, and the hands-on process of making the cider. As one marketing friend said, 'You've nailed it. You're selling them the dream.'

It had certainly taken us long enough to get the formula right. Every smallholding is unique, with location, climate, size and the skills of the smallholders all playing significant parts in the success or failure of their projects. We had a rough start, which continued for a few years, until fate and common sense pointed us onto the path we were then firmly on. We had made a loss for a good few years and were finally making a profit, although working incredibly hard to do so, and this was not what we had planned when we moved from Scotland. Thankful for all the successes we'd had recently, I was still watchful for the Enforcement Officer and never took for granted that we were extremely lucky he had never reappeared to question what we were doing; although technically the biggest earner was the cider and, technically, this was still our main business on our smallholding. It was like having an ominous black cloud permanently on the horizon.

As the hedges started to sag with the wild harvest, I took the opportunity to make my usual few batches of sloe vodka, some bullace vodka, and then made jars and jars of pickled beetroot from our vegetable patch. The cupboard soon swelled with jars of jewel-like colours and inevitably, some demijohns of bubbling country wines.

****

Damson or Sloe Vodka

1 lb damsons or sloes. You can either prick them all over or pop into a freezer bag and freeze for a week, which on thawing, breaks the skins.
4 oz sugar
Quantity of vodka (or gin if you prefer)

Place the fruit in a large screw top bottle with the sugar and cover with the alcohol. Seal and shake vigorously to dissolve the sugar. Leave in a cool, dark cupboard for 3 months, shaking vigorously once a week. Strain and rebottle. You may add a few blanched almonds if you like.

****

As the nights drew in, and autumn leaves swirled around the smallholding, trying to gain entry through doors, we snuggled up in front of the wood burner and talked about expansion, retirement and the direction we were headed. The garden machinery business had become very successful and even though John missed the teaching, he had continuous machinery work from January through till October. Then he helped me with the cider production till December. But we were feeling our age, and as we badly needed a rest, we planned another escape for the following March, before the machinery season started again. The thought of wall-to-wall sunshine cheered us through the dark, short days. Cornwall, with its south western location can be very damp and wet and this winter was proving frustrating to us both. My old winter cough had returned and I was barking like The Hound of the Baskervilles, and consequently sleeping very little. Our GP refused to offer any medication, stating that it was the damp atmosphere and to try to get away for a break in the sun! I had to replenish the stores on the three beehives twice that winter, as it was mild enough for the bees to be out of torpor, but not dry enough for them to be out foraging for feed. I had serious concerns that the weakest colony might not live out the winter, but it was a horrid game of wait and see.

I missed the college and the Rural Business School and tried to keep in touch by phone, but our time had passed and I was left with good memories. I decided to write again and completed a couple more smallholding articles, but never sent them to the magazines. My heart was not in it, and I felt incredibly tired and empty. The staff at the nursing home had told me it could take up to two years to get over the death of a loved one, and I realised just how intertwined and complex lives were, wondering if I would ever be free of my mother's influence.

Just before Christmas I attended the dinner and awards ceremony for the Cornish Sustainability Awards, seeing this as a good opportunity to network and meet like-minded people. It was bustling with business people from all walks of life from Cornwall and beyond, all trying to make a difference in the field of sustainable practice. I felt excited and a little overwhelmed to meet so many people like-minded people making small and big changes in the way they ran their businesses, and my head was filling with ideas that I could implement to improve our own practices. As the awards started to be called out, I was flabbergasted to hear *my* name called out as joint winner in the Best Individual category. I sat, dumbstruck in my seat, as applause rang out, and the lady sitting next to me urged me up to the stage to receive my award. It was heart-warming and humbling to think our methods and lifestyle, our silly sustainable practices that were so important to us, and yet so small in the bigger picture, were of great interest to other people too. Everything happened so fast, I really couldn't take it in at the time, but when I came home with the award, my husband's face was proud and smiling, and my son was delighted and proud of his mum.

Kieran celebrated New Year by getting completely and comprehensively drunk and was brought home almost unconscious by friends at the pub. He arrived with very little dignity on the back of a flat bed truck, stinking of spirits, stale beer and vomit. We've all been there as teenagers and I was cheerfully told that he had been profusely sick already and should be fine to sleep it off. The lads dragged him inside and deposited him on the kitchen floor in front of the wood burner. I found a blanket, put him in the recovery position and settled myself in the old armchair to watch over him, feeding the wood burner every so often to keep us both warm. At five am I managed to help him to bed, thinking he was over the worst, and then headed stiff and cold to my own bed for a few hours. He weepily apologised when he woke, had a shower and courageously headed back to the pub for his lunchtime shift.

Jaimie was still playing tricks on us, and after having carelessly chucked off my boots one evening on the porch, I headed for a bath and an early night. Next morning, I automatically grabbed the boots and pulled them on. Instantly feeling something wet and squidgy under one foot, I recoiled with a scream and threw the boot off. My sock was quickly pulled off and examined and was wet with blood and flesh. Gingerly feeling inside the boot, my hand alighted on the very mushy remains of one of Jaimie's "little offerings"; an almost unidentifiable and now flat-packed rodent. Of course, we were used to our great mousers doing their job in keeping the numbers of mice down, but in this case, one of the dying creatures had sought a last refuge in the hastily discarded boot. After this disgusting episode, John fabricated a welly rack for outside the door where we could safely place our upturned boots to prevent a recurrence.

All the previous autumn's cider was bottled and some labelled. We had created six individual batches and the printers had created six wonderful labels for us. Farmhouse, Quencher, Elvenny, Gold, Reserve and Seadog.

Each one was unique, because of either the mix of varieties or the way it was produced – either still or sparkling. This was our best year, and although hard work, we were pleased with what we had achieved. Just over 4500 litres of Spotty Dog Cider were ready for sale and we were enjoying a little fame in the county and further afield. Labelling had to be completed by the end of March, before our precious few days holiday, and we pulled all the stops out to ensure we were ready. The cider shed was stuffed full of boxes of bottles and trays of bag-in-box bags. Empty barrels had been washed and stored and the cider equipment was packed away for another season, and we should have been able to relax, had we not heard some disquieting rumours.

Cornwall is a small county and, in the farming community jungle drums beat fast and loud, and I soon heard about some smallholders who had fallen foul of the planning system. Although we had experienced no further contact from anyone at planning including the Enforcement Officer, I was still wary and nervous. Although we were technically still working the smallholding, the majority of our income came from cider production and garden machinery servicing. According to the strict wording of the Agricultural Occupancy Condition, neither was fulfilling the tie, as machinery servicing was a service and cider production could be construed as either ancillary food production (if the majority of apples were grown on our holding) or food manufacturing (if mainly bought in).

I contacted a few friends who lived further up country who had contacts who were living with or had lifted Occupancy Restrictions and started to tentatively gather information. Through all these contacts, I finally heard about a family in Devon who had bought some land and had finally, after years of uncertainty overcome planning issues, gained permission to live on their land and build their home there.

Now, I have to say their situation was very different to mine, but the material point was that the lady of the family had managed to retain the services of an ex- council Enforcement Officer who had set up as a rural planning expert. I got in touch with this lady, who immediately understood my hesitation at contacting anyone official. However, she put my mind at rest with a brief discussion about the merits of the man in question.

Taking a deep breath, I telephoned his office and made an appointment to have a brief discussion about our situation, reasoning with myself that, after all, all information is power.

John and I had a business lunch in our local pub to discuss the future of the cider making, and agreed not to expand any further. Money wasn't everything. We would instead get volunteer helpers at the start of harvesting time, and try to manage our time better. I think we both knew that we were putting off the inevitable conclusion that we were struggling physically with the success of our business, but this was small fry compared to the subject I then aired.

Towards the end of our lunch, I mentioned that I had made an appointment to discuss the Occupancy Restriction with someone out of county who had previously been an Enforcement Officer. John went quiet. For the last nine years we had managed to avoid any planning entanglements and, although he understood that a visit could come at any time, he wanted to let sleeping dogs lie. The meeting was arranged for the the start of April. I reassured him it was merely a fact finding mission and I would be divulging very little. When we returned home, I prepared a little file with the details of the wording of our Occupancy Restriction and some notes about methods used to remove the restriction and waited for the day of the meeting.

9.
Opportunity Knocks

Our short holiday in Dordogne was delightful, and we explored the perigord noir; an area of outstanding wild beauty, kept green by many rivers and streams in the area and with small, mixed farms spread over large tracts of land. Walnut groves and sunflower fields were nestled between soaring limestone cliffs and rubbed shoulders with chateaux along the valleys. When we returned, I collected two lambs from the farm and installed them as usual in our paddock. The joy of having them was better than all the anti-depressants and steroids. Just watching them play and run was soothing to the soul, and I spent many free moments with a cup of coffee watching them.

We advertised for volunteers on a farm helper site, called HelpX and had lots of youngsters coming from all over the world to help us in exchange for bed and board. Most worked hard, were keen to learn about cider making or bee keeping, the chance to improve their English and loved exploring the beautiful coastline and Cornish heritage. We learned a lot from these visitors too, many of whom were great fun.

Cider sales began again at the start of Easter and I returned to having a stall at the Food Fair in Padstow. Padstow is a pretty town, even in the height of tourist season. Mornings are fresh and crisp, my favourite time of day. Once I'd set up my stall and arranged all my bottles, I'd walk round the harbour, grab a bacon butty from my favourite take-away and stand for a few minutes watching the fishing boats head out to sea, before heading back to the Food Fair. Then we had a steady stream of visitors and sales all afternoon till five pm. Every evening I returned home with empty boxes and most of the stock sold. People enjoying the cider, and asked about our orchard and process and this was particularly rewarding after all the hard work producing it.

I returned late one afternoon to find the smallholding very quiet. I headed for John's shed to find him sitting at his workbench, rebuilding a lawnmower gearbox. 'Something's wrong. It's too quiet,' I said, and walked round to the back paddock.

At first glance I couldn't actually see anything wrong. It was unnaturally quiet. Then I saw huge smears of blood on Kieran's bedroom window and a pile of feathers at the foot of the house. I walked over to find the remains of a beautiful Buff Orpington hen. Quickly casting my eyes around, I saw another bundle at the fence.

'John!' I called loudly. 'Bring your shotgun!'

Fearing we had been visited by a fox, my first instinct was to try to find the rest of the birds. At this time we had ten pens all with rare breeds in breeding sets; in all around 90 birds.

I soon found Charlie, our beautiful and very tame Buff Orpington cockerel. He was dead and it looked as if it had been mercifully quick. He had been one of my favourites, a friendly, gentle giant who loved to be cuddled and fussed. Tears streamed down my face as John hurried around the side of the house looking grim.

'I've found some alive in the trees, you'd better come and sort them out.'

Meanwhile he headed towards the small gap at the bottom of the fence where another two piles of feathers lay, this time black ones. In total we had lost all seven of my beautiful lavender Orpingtons, including the cockerel, and another five Orpingtons of various colours. The only surviving live birds were the Light Sussex, and a couple of younger Orpingtons, who had managed to scramble up into the trees and bushes. Piles of feathers in the paddock floated up in the breeze. John hadn't heard anything from his work in the shed of this killing frenzy.

John grimly collected the dead birds in a feed sack and walked round the perimeter. Then he saw a quick black movement in the Miscanthus and out popped a small black poodle! He called me over and pointed. A moment later a little fat Jack Russell came out the field stained with blood around its mouth, wagging its stupid tail.

'Isn't that the Jack Russell from the farm in the next lane?' he asked.

'I don't know, but it's a hell of a coincidence if it's walking round our boundary with its mouth all bloody and we've just lost so many hens!' I fumed. 'Shoot them!'

'I can't shoot them! They haven't been caught in the act!' he growled at me.

I pulled out my phone and managed to photograph them both. Then I photographed Kieran's blood-smeared window, the blood obviously from an injured hen trying to escape the dogs. And then I phoned the police. They reluctantly turned up the next day, and duly noted my complaint and anger, saw the bags of dead hens and the photographs. I didn't expect to hear any more from them, but a few days later, the officer returned to say he'd traced the dogs to a nearby farm and that they were prepared to compensate me for my loss. Although still angry, what could I do? The dog owners had agreed to compensate me and keep the dogs under control in the future. John informed the police officer that if we had any further visits the dogs would be shot instantly.

The holding was very quiet now. We moved the Light Sussex hens round to a safer pen, and found three more blue Orpingtons in the bushes and one terrified lavender hen which we housed together in a house with an enclosed run. John offered to come with me to buy some new Orpington's from a breeder in Somerset, but I couldn't even discuss it, I was so upset. I'd bred all these birds myself, and took the mass killing very hard, knowing it could so easily happen again. In fact, I was so upset, I decided there and then to re-home the surviving birds to ensure that we wouldn't have to ever see a repeat of such carnage.

I left early on the Friday morning to drive to Cullompton for my meeting with the Ex-Enforcer, as I had started to refer to him. Apart from John, I hadn't told anyone about this meeting, and John warned me not to actually give the name of our place until I was sure I could trust him; as if I would! The last thing we wanted was have word get back to Cornwall Council that we were looking at the possibility of lifting the tie. Careless talk costs lives, as they say and as we were so close to having done ten years on the holding, and still didn't have enough information about what our options were, this was the last thing we needed.

The office was tiny and a chubby black Labrador ambled in to say hello to me in the waiting room, followed by a stocky man with a beard and a tattersall shirt and brown cords. Glenn offered me tea, which I declined and I took a deep breath and started the ball rolling.

'I've been told I can trust you, but I'm sure you'll understand why I don't want to tell you exactly where I live or the name of our property.'

'Well, can I ask how you heard about me and what you think I can help you with? If we start there, it's as good a place as any.' Glenn said reasonably.

So, I mentioned my informant and that she had suggested

a meeting to discuss if we could lift my Agricultural Occupancy Condition. I also mentioned I knew a bit about the case he had helped her with and that if she felt he was a trustworthy man; then he was probably the person I needed to talk to.

'Can I make some notes?' he asked and reached for a pen. He noted my hesitation and smiled and put the pen down.

'Nothing you say here will leave this office. You can trust in my complete confidentiality.'

'It's just.... Well, we're *so* near to the end of our ten years, but I'm so worried...'

'Do you have a copy of the wording of the tie?'

I passed the typed-out copy over to him and waited till he had read it through.

He asked a few questions which I answered as truthfully as I could without actually naming the house. Soon this short exchange was done. Glenn wrote a few notes, pushed the piece of paper towards me and sat back. It only had a couple of words and dates on it, but I had told him very little, so I wasn't exactly surprised.

'Now - you've told me the one and only visit from the Enforcement Officer was in July 2006, and, after you sent him a follow-up letter outlining what you intended to do as a business, you have had no further communication from him or the office?'

'That's right. That's to say - he may have visited when we were out at some time, but no, nothing, no letters or anything.'

'And you ceased the nursery business...?'

'Well, it was just never making money. I kept the books going for the first year and then phoned the tax office to explain the situation. They were only interested in the fact that I was still trading. The business had changed, and we had no money...so John started working and then I started teaching…but we were still working on the smallholding and running it at a loss.'

I gave him copies of my tax returns, and explained about the failure of the business, about teaching and John's employment, then about Mum coming to stay and how this really affected what we were doing. Eventually the whole story of what we had done in the last ten years just spilled out.

'We just need to know what we can do now,' I blurted. 'It's just so tiring, and we're getting older. I don't want the shadow of Enforcement hanging over me any more. I can't do it any longer. I've read that we can put the house up for sale and see if we can avoid any interest in it for two years, but that's not going to happen. Someone will buy it. It's a nice house now and in a great spot.'

I reluctantly pushed a photo of our house towards to him. I'd spilt the beans now, so I might as well go the whole way.

'But this is a *charming* house!' Glenn smiled broadly. 'Of *course* you don't want to try the marketing approach; that's not the right decision to make in your circumstances.'

'I don't know what to do anymore,' I whispered, tears finally escaping and splashing on the table. The relief of finally being able to tell someone about all the years of worrying all came out in one uncontrollable deluge, and the poor man handed me a box of tissues.

'Oh no! Don't be nice!' I wailed, 'or I'll not be able to stop!' and the tears kept spilling down my face and over the table.

'I'll get a new pot of coffee,' he said, patting my shoulder and tactfully leaving me for a few minutes to blub like a kid and try and pull myself together.

When he returned and I'd had a very sweet black coffee, we sat and had a chat about what we did. I told him about Mum, and the cider and the fact that we were struggling physically to keep the smallholding going.

'Mrs Turnbull, although you bought the house with the best of intentions and indeed tried running the property in strict accordance with the AOC, you've been breaking the conditions of the Agricultural Occupancy for some time. By the end of July you'll have done your ten years. Opportunity is knocking. Are you up to opening the door to it? I'm delighted to tell you that, in my professional opinion, we can get you a Certificate of Lawful Use for your property, which means the Enforcement Officer *won't* visit, and that you can either continue to run whatever business you currently run, or you can proceed to sell the house at full market value.'

I sat stunned for a few seconds. Part of me was delighted and relieved and the other part still terrified that he had somehow misunderstood our situation, and it would all go terribly wrong.

'We need to wait to put in an application for the CLEUD till after the tenth anniversary of the date of the letter you sent to the Enforcement Officer - say a month after? Meantime we have lots of time to collect evidence to support what you've told me. I'm sure, as a business woman, you've kept all your paperwork?'

I was still in shock, but nodded.

'Tell me again….it's *definitely* all going to be okay? We can get a CLEUD and then sell the house if we want?'

'Yes, or stay there and continue as you are now. It's up to you, but there will be no more Enforcement visits.' He said smiling.

'And you're 100% positive? Because once they know...' I ended ominously.

He nodded. 'Yes, it'is a strong case and it really will be fine.'

He paused a moment to let this sink in.

'I'll send you an email listing what sort of evidence you'll need to build a strong case.'

'And, er...how much will this all cost?'

He wrote the sum on the bottom of the notepaper.

'This is my fee and includes the cost of paying for the application. Your case is a very simple one. Please stop worrying.' *Cheaper than the cost of a holiday*, I said in my head.

Leaving the office, my heart thumping in my chest and tears a few minutes away again, I realised we had reached a turning point. It really looked possible we could actually free ourselves from the threat of Enforcement, and also from the drudgery of having to continuously be seen to be working our tiny one acre smallholding. The relief was huge as this could free us to change how we lived. If that indeed is what we would decide to do.

One thing was for certain. I had to start combing through old files and paperwork to amass the paperwork required to build an iron clad case to get the CLEUD. This was absolutely crucial and would take time to collect, and we couldn't relax till this was done and the application was granted.

Glenn would take care of the application itself, maps and the technical stuff. I had to organise having some Statutory Declarations done as a starter. Glenn emailed me a blank one, and John and I completed one each as the primary declarants. Basically a Statutory declaration is a legal document where you set down the terms of something. You then sign this in front of a solicitor who witnesses it as a sworn declaration, and countersigns it. To be absolutely sure of the smooth acceptance of our application, I also drew one up for our neighbour Ron, our friend Colin, and our retired police friend, David. I spoke to them individually and thankfully, they were all happy to help, and we arranged a visit to our solicitor the following week, when we could all swear and sign our respective copies.

So far, so good. I then had to ask another couple of friends to support our application by writing letters to the planning department, confirming the failure of the plant nursery business, and some other matters pertinent to the wording of the agricultural tie. Again, I gave them an example of what to write, told them what NOT to say, and asked to see a copy before they posted it. Once more, this went like clockwork.

The hardest thing was completing an exact career record for the whole ten years for both John and myself, mainly as, in that time, one place of work had gone into liquidation, one said they didn't keep any employee records as far back as that (not that I actually believed this), but luckily I had kept end of tax year paperwork, and payslips for the whole period. This all went into a separate folder as there was so much and it took ages to get replies back from various offices.

Whilst all this was going on, I took my eye off the ball, mis-timed checking the beehives and saw we had a hive with capped queen cells in it. The hive was bursting with around 50,000 bees and I had not seen the queen cell until it was clear it was capped. This meant that any time the older queen would fly off with a gang of bees to form a new colony. As I didn't want to lose them, I quickly went to the shed to make up a new hive. So, base, brood box, queen excluder, super and roof were assembled. I had to make new frames with wax foundation for the brood chamber, and this fiddly job took a little time. I hurried to grab a sandwich for lunch and John came in, looking for a cup of tea. 'There's a few bees in the shed,' he remarked.

'Yes, I'm assembling a new hive to move the bees before they swarm. There's at least one capped queen cell in the hive near the orchard,' I replied.

Sadly, when we returned after this incredibly brief break, it was clear the bees had found the new "home" and couldn't wait to move in. I'd left the shed door open as it was warm inside, and the smell of honey and wax was irresistible to them. The colony had indeed swarmed and instead of us having to track them down, they had helpfully moved straight in! The shed was full of flying bees and the half-completed brood box was full of happy, nosey bees.

I returned to the house, popped on my bee-suit, returned to insert the remaining frames into the brood box, put the lid on and moved the hive outside onto a stand outside the shed. The remainder of flying bees would go in as the evening approached and we would simply move the new colony to a stand near the existing hives. A sprig of leylandii draped in front of the entrance of the new hive would force the new colony to reset their "bee navigation system" and imprint the new location as home. I would return to both hives in a few days and check them, and hope the original colony would soon have a new hatching queen who would go on a romantic mating flight, and come back ready to lay thousands of eggs to populate the new hive.

The maps for the application to remove the tie were duly drawn up. I checked them, they were altered slightly as they weren't perfect, and the application was soon ready to be posted. We spent an afternoon checking everything at Glenn's office and that was it, in the post, recorded delivery. Then about a fortnight later, I returned from feeding ducks to a message on the phone. The Planning Department from Truro wanted me to return their call to arrange a visit to the property. I immediately panicked, thinking it's an Enforcement Officer wanting to visit, and phoned Glenn in tears.

He calmed me down eventually, found the name and phone number of the officer from me and told me to go get a cuppa and he'd find out what the story is. A few minutes later he called to say it's a standard visit they have to complete to verify people are actually living in the house.

'Don't panic,' he soothed, 'this is normal. He only wants to come and take a couple of photos for the file. It's perfectly normal.'

'Take photos of what?' I demanded defensively.

'Of the house – to prove it's a house. Of the kitchen or the living room – to prove it's a dwelling with people actually living there.'

I calmed down.

'It's just for the file. Perfectly ok. I actually know the officer slightly. He's nice. Ive explained you're very nervous and that you've had a bereavement. He'll only be there ten minutes. He doesn't care that you're putting in the application. Ten minutes. Think of the Big Picture. This is the last piece of the puzzle.'

I took a deep breath and forced myself to calm down. The last piece of the puzzle. *I can do this*, I say to myself. *It's a ten minute visit.* Next Friday at 3pm, Glenn says.

'Write it down now', he repeats.

I wrote it down. My head was up my backside and I couldn't concentrate with the pressure of it all.

The days dragged past. Lack of sleep meant I was irritable and anxious. I checked and rechecked the files I have sent to Glenn. I knew I had to be patient and wait for the slow grinding wheels of bureaucracy to turn. It was fine. But by Friday I had suitcases under my eyes. It was really not fine, and I waited, waited, waited for some unforeseen problem or rule that I didn't know about to be brandished in front of me. It was all going to go pear-shaped. Kieran and John avoided me and trod on eggshells.

Friday arrives and I try to keep busy. A minute lasts an hour. At last a strange car drives into the driveway and a young man appears with a file and a camera. Lily bounds out to see who it is and he pats her enthusiastically. *He must be nice*, I tell myself, *the dog likes him.* He apologises and introduces himself. He'll leave me in peace as soon as possible and realises how stressful this all is. He'd like to photograph the front of the house for the file.

I nod, and my throat is too dry to speak.

'And the kitchen if possible?' he asks. I lead him inside. John is pretending to read the newspaper in the kitchen. He looks up, shakes the guy's hand and sits back down, pointedly ignoring me. The photograph is taken.

'And a bedroom?'

I lead him down the hall, open my bedroom door and realise the bed isn't made. I close the bedroom door and lead him to Kieran's room. I open the door and the words,

'Jesus Christ!' escape from my mouth.

It's like a bomb has exploded in there. Clothes and music CD's all over the floor. X Box games littering the place mixed with cables. I mentally roll my eyes and wish I'd thought about a quick tidy up. There are even a few empty cider bottles lying on the floor. Andy, the Planning Officer laughs.

'Oh, I have two boys!' he says smiling and taking the required photo.

Mission accomplished, I walk him out to his car. It's over. He shakes my hand and takes a look around. The sun is shining on the Camel Estuary in the background. The hills towards St Kew are green and beautiful and the Miscanthus is almost waist high, swishing joyfully in the breeze. I can hear a skylark over beside the hill fort and the air is heavy with the coconut smell of a million gorse flowers.

'It's a fantastic spot, Mrs Turnbull. I can't imagine you'd ever want to leave, but I'm approving the application and you should get the official paperwork from my office in a week. I do wish you well whatever you choose to do in future.'

He smiles and is gone, waving out the car window as he turns down the lane.

John is standing by my side and silently reaches for my hand.

'All ok?' he asks gently.

I nod, turn and disintegrate into his chest, blubbing like a bloody fool.

*****

We had done the impossible. We had taken a virgin one-acre piece of land and created a wonderful smallholding. We both now had good businesses working from home. John's garden machinery business was busy all year with a short break when he helped me with the cider making; and after a lousy start with the plant nursery, I had created a wonderful, respected small cider business. The children had grown up in a delightful setting, far away from the temptations of urban life in Glasgow. As a family, we ate well, with our own lamb, chicken, duck, eggs and fruit and vegetables. We bartered some lamb for pheasant or partridge, and the spent apple pomace in the autumn was swopped for pork and sausages arriving at Christmas. It had been hard, hard work, but we were now looking at an opportunity that would allow us to have a choice of what we could do in the future. Although I didn't have the certificate in my hands yet, my heart soared with relief.

*****

Of course, life goes on and there are jobs to be done. Bees have to be checked as usual and I don my suit one morning and head to the hives. All is good; the bees are happily gathering pollen for the baby bees and turning nectar into honey. I return to the house and with the door ajar, strip off my bee-suit. It's very warm. The smell of the honey and bees is around me and before I know what is happening Lily, our nosey Cocker Spaniel has dashed past my legs and is following my scent to see where I've been. I call her sharply, but it's too late.

She has run round to the hives, and the bees, affronted at the sudden intrusion into "their" space have attacked. Lily is yelping and I run round in a panic to find her writhing on the ground in a cloud of bees. I pull her away and brush bees from her head. A few have stung her and she yelps and cries. I carry her back to the house and try to calm her, but she is terrified and obviously in pain. An examination reveals a number of dead bees on her face and near her eyes. I pull them and try and remove the embedded stings, noticing with horror that she appears to have been stung in the eye itself. There is nothing

I can do except drive her to the vets quickly, and so she is bundled into the car and I drive to Camelford to our local vet.

A special dye is squirted into her swelling eye, and a thorough examination is made.

'Yes, I can see the sting which is still embedded in there, says our vet.

'I'll have to operate to try and save the eye,' he ends, stoking Lily's head. Poor Lily has been very good sitting still and obviously in pain. My heart sinks.

'I can't operate here. Can you take her to the surgery at Bude? I'll phone ahead and get them to prepare for surgery. I'll drive up now, if you want to follow me.'

I cry and know I need to pull myself together to drive the forty minute journey. I return to the car and phone John to update him. The chances of saving Lily's eye are around fifty-fifty. An hour later I hand her over and I wearily drive home. The operation will take at least an hour and then she has to recover from the anaesthetic. John is very quiet.

Whilst I've been away he has fenced off the hives to prevent a recurrence of this horrible incident. We have a horrible long wait and at teatime we both drive up to collect her. She has a patch over her eye and is wearing a big plastic cone to protect her. The vet thinks he has saved the eye, but is unsure at this time if she will have sight in it. We have to wait for a month to allow the surgery to heal and see if the grafting operation to cover the hole cut into her eye to remove the sting has been successful. I drive and John cradles her on his knee the long journey home. That night, Lily is forced to endure short toilet trips to the garden on a lead, and is now sleeping on our bed.

The weeks pass and Lily eventually gets her plastic cone off. The vet is pleased and to our great relief, she can see. However, this harrowing episode has forced me reconsider keeping bees. I advertise the colonies separately and they are bought quickly, as are the bee-suits and all the assorted bee equipment.

We now have no poultry, no ducks and no bees. Our smallholding is almost bereft of livestock, and the dynamics have changed. It's hard to identify the many conflicting feelings and emotions we have at this time.

Everything has changed. I feel like a sailboat with no rudder, and even more changes are in the wind.

Kieran accepts the offer of a place at Manchester University, and packs an enormous amount of stuff he wanted to take with him. I arrange to drive him there in our estate car and I'm dreading it. I know this is the culmination of all those years of bringing up a child; of preparing him for the big world. But we are incredibly close and I can't bear it. I know he will return at Christmas, but he will never be my "little boy" again. The house will be too big and too empty and too quiet. Apart from his lovely girlfriend, there is little in Cornwall to bring him back. We will soon be empty nesters and I feel very old.

The week before he leaves the Certificate arrives. It's

official. The relief I feel manifests itself in a fit of crying. I feel like I've been holding my breath for over 10 years and can finally breathe freely. We email Glenn and thank him. Words don't explain our gratitude and seem trite and small for what has been the most significant change for us. Finally the crushing pressure that the Agricultural Occupancy put on us was officially gone. We were free agents. John insists we go out for a celebratory meal and we use this as excuse to take Kieran as his "farewell send-off".

10.
End of the road

At night we whisper together. I hardly sleep, and unfortunately, because of this neither does John. In the morning we decide to talk about "the future" once I've returned from taking Kieran to Manchester. And so, all the vast amount of stuff is squeezed into the car and I wait for Kieran to say goodbye to John. There are tears and fierce hugs and then my grown-up boy says, 'Let's get this show on the road, Mum.'

It's a six hour long drive and I'm treasuring every second. We drive, play music, sing along to songs and have the most wonderful conversations about bands that have influenced us, and friends of Kieran who are moving in different directions. His girlfriend is staying in Cornwall, but has applied for university in Manchester the following year. She's pretty, clever, sensible and loves Kieran to bits. John has a real soft spot for her and I am happy for Kieran. As a couple, they have to try the long distance relationship and see how it pans out. He confides in me with a laugh how they have "somehow" broken the slats in the bottom of his bed and there is a plastic crate holding it up. I have to laugh, though my heart is breaking. My little boy is grown up.

All too soon we have arrived. I unpack the car, whilst he runs up and down stairs with his belongings. He's excited and eager for me to leave. My heart is like a stone in my chest, but I take a deep breath and tell him I'll get a parking ticket if I linger. One hug and I swallow the tears, forcing them not to come until I've driven away. I open my hands. This little fledgling takes flight, and waves with a final cheery grin before darting back into the building. I start crying around the corner and cry all the way home.

*****

It's so quiet in our home that it's unnerving. John and I no longer have to taxi Kieran to work or to college or to rugby. The electric bill drops as twice daily showers and copious laundry ceases. Kieran phones to tell us he's loving Manchester and is eating a lot of rice meals. It seems his student flat in Manchester has five Chinese girls resident, and they make the food in exchange for Kieran doing the washing up. At least I know he's eating. John and I visit our favourite coastal spots but they fail to lift my spirits. The smallholding is quiet too. I have nothing to do and have decided not to make any cider. John has loads of machinery still coming in to service and leaves me to my thoughts. To deal with the boredom, I write an article about smallholding profitability, which grows and grows. I've written loads of articles before, so how much harder can it be to write a book?

Together we wander into Wadebridge one evening and have a drink in one of the pubs. John hesitantly talks about the possibility of selling up and moving back to Scotland, to Oban, to be exact.

'Oban? *Oban?* Where it's dark five months of the year and full of midges for the other seven?' I exclaim, shaking my head.

'We could retire. Properly retire,' he says quietly.

I look at him. He's 59. Yes, of course he's thinking of retiring. If not now, when?

'How much do you think we could get for the house?' I ask. 'I know its worth at least a third more now that it is free of the tie. Bit of land, great views down the estuary, just a mile outside Wadebridge…'

'Why don't you have it valued, and we'll take it from there,' he suggests.

So the next day, I'm in Wadebridge organising valuations with estate agents. Within a week we have a price. It's a good amount and we both know it will sell quickly. I don't want to go to Oban. John searches the internet for rural houses down the West Coast of Scotland. I'm not keen. The weather puts me off. The midges put me off. I feel that returning to Scotland would be a backward step. Back to awful weather and winter darkness. We've done that - and I don't want to return to it. I suggest we go away for a week, now we don't have animals or poultry. So we organise a friend to come and stay with the dogs and we fly to Bergerac again.

In a few hours we are sitting outside in the sun on a restaurant terrace, soaking up the heat and quenching it with local wine. I lightly suggest the possibility of moving here. Yes, it's a huge move, but we made a huge move from Scotland to Cornwall and survived. Are we ready for a new challenge? Are we too old? Other people have moved from the UK to France. Thousands. Houses are slightly cheaper, weather considerably better. We look in estate agents windows (immobilieres) and gather information. We speak to ex-pat Brits living in France and the small off-hand idea grows into a real possibility. What is the saying? A change is as good as a rest? Maybe that's all we needed.

Back home we return to our local pub and wax lyrical about the weather in France. Gossip and news travels fast round Cornwall's grapevine. Locals are aware we have had the agricultural tie lifted and the expectation is that we will sell up soon. In fact, we are approached by a slight acquaintance from the pub. He has just sold his father's house and is interested. I'm sceptical, as he has previously "impressed" me as a man full of talk with little substance, but I merely say we're not yet on the market. At home, we empty the kid's rooms, carefully storing all personal things. I start to collect cardboard boxes and John sorts through his shed. We have a lot of stuff to sort through so are starting even before the house is officially on the market. In a week or so the estate agent comes round, we prepare a schedule and photographs are taken.

The man from the pub, (we'll call him *All Talk* for reasons that will soon become obvious) appears when I'm out one day and asks John for a walk round. He is standing in the yard beside his land rover when I return. He's interested, he says, and I realise he's been given a tour in my absence. I give John "the look", and he has the grace to look uncomfortable so, as the ball is now at my feet I take the initiative and give *All Talk* the hard sell.

'It's on the open market and I'm firm about the price. Please don't bother if you want to offer lower as it won't happen. We know it'll sell quickly.'

I head into the house and question my gut feeling, which is usually pretty sound. How he can afford our smallholding is beyond me and I'm cross at John for showing him round without any notice or prior warning. As he also comes into the kitchen I see that he's picked up on my mood.

'I'm sorry. He just said a quick look and then, well…'

A week later *All Talk* phones John to see if his partner can come up. I tidy and reluctantly agree. I don't like either of them and wonder if this is just nosiness, and resign myself to the possibility of a few such visits from nosy locals. After the tour of the house I ask how they can afford the price. I know he has no job and she works part-time. They have sold his late father's house and are selling her house and her mother's house. Are selling. Not sold. Again, I explain I'm looking for a firm mortgage offer or cleared funds. Something just doesn't feel right here, but I can't dismiss them out of hand. After all, even if I don't like them, I'm hardly going to be living next door.

Meantime, we have decided to have a look at South Eastern Dordogne area and head off for a few days. The dogs are off to their holidays to kennels for a couple of days and we have a delightful time driving around the Vezere Valley. I even drive a little, and once I get used to the gear stick being on the wrong side, I'm fine and driving confidently. We love the area. It's like Scotland with sunshine. Pretty golden-stone villages with clay tiled roofs, hills, rivers and towering limestone hills. So we start house-hunting in earnest. Some are too big, some too small, others need loads of renovation and some have too much land. We keep our feet firmly on the ground and shortlist a few. But only one is shouting, 'Buy me, Buy me!', and we hesitate as we haven't had a firm offer on our Cornish house.

Back home, *All Talk* makes us an offer which we reluctantly accept, after being assured by our solicitor that their solicitor has seen they have cleared funds. The process of searching will take a while, so whilst John clears his servicing backlog, I search the internet for possible properties and research the locations. We tell the children, who are a little shocked at both the speed and the decision to move to France. Kate is a little incredulous and reminds us we don't speak French. Kieran quietly says, 'If it's what you want Mum, then go for it. You wouldn't tell me not to do what I wanted, so I can't say anything against it.'

I put in an offer on the house we liked in France with the proviso that we can back out if our house sale falls through, gather boxes and pack. John sells loads of equipment and machinery, but I'm still concerned at the amount he wants to keep. He, in turn is concerned at the amount of books I want to keep.

About three weeks into the search time, the lawyer is concerned she still hasn't had confirmation of funds from the buyers. They also delay with small things and I'm growing concerned. By 6pm on the date we are due to exchange contracts they fail to sign. My solicitor phones and tells me. She has had trouble contacting their solicitor. We're both worried and I'm furious.

All Talk phones our house after six, trying to explain that nothing is wrong, they just have a little delay.

'You don't have the money, do you?' I say angrily.

He blusters a little but fails to convince me we will *ever* see the money. I hang up. To say I'm furious is an understatement. My solicitor calls me at 7pm. She has finally got hold of his solicitor who confirms they cannot proceed, due to lack of funds.

'I thought their solicitor was supposed to check this?' I accused, and before I hang up add the insult that this would never happen in Scotland.

Trying hard to control my fury, tears and heartbeat, I walk to the bedroom, find the little bottle of steroids from my bedside drawer and take one. I know the last thing I need right now is a racing heart or heart attack.

The next day *All Talk* tries to call me and again, and again I hang up. I'm now waiting for official confirmation that the sale is void. *Thank God we haven't signed anything in France*, I think. I'm so angry I can't speak to John; which, on reflection is unfair. He's only tried to consider every possible offer, and is furious too. When we visit the pub that evening the jungle drums have been working overtime and, of course, its common knowledge what has happened. John jokes to me that the village was probably aware he wasn't going to sign the contract before we knew! It's a joke in bad taste. Even worse, as we are sitting there trying to distract ourselves, *All Talk* sidles up to us at the bar to try to talk to us. The bar is suddenly silent and alert. I look round to see the reason for the silence and see red.

'You have some nerve!' I shout.

'Can I have a quiet word…?' he says under his breath, but the whole pub has stopped to listen.

'You can go and take a flying fuck to yourself! You *knew* all that time you didn't have the money!'

'No, we *have* the money…'

'NO, you *don't!* Our lawyer told us that *your* lawyer told them you *don't* have the money! I wouldn't sell my house to you now if you were the last person alive! Don't you dare speak to me. Don't phone me again. We're done.'

I turn my back and down my glass of wine in one. Well, that little scene has provided the village gossips enough to dine on for a few months. I order us both new drinks, refusing to leave. After all, *we* haven't done anything wrong. Some time later, when chatter returns in the bar, an older friend from the village quietly approaches me to say they've known all along that he didn't have the money.

'Why didn't you tell me?'

'I didn't know what to say,' she admits awkwardly. Apparently most of the village was aware.

The house went back on the market on the Monday, so I telephoned the immobiliere in France to explain the situation, and withdrew our offer. So far, I'd incurred fees to pay in France and solicitors' fees in Cornwall too, through the casual behaviour of a daydreamer and liar. We had lost the house we liked, but thankfully had not lost the ten per cent non-returnable deposit. John mopped up a few more repair jobs and I wrote and edited more of my book. Kieran would visit us at Christmas and stay at his girlfriend's wonderful parents, which was a huge help.

In the run up to Christmas we decided to have a last dinner in a pub in the village of St Teath. The owners take Christmas seriously and have the unusual habit of lining the walls and ceilings of the pool room, the restaurant and the bar with silver foil. Over the years we had seen this many times, but it would (hopefully) be our last experience of having some idea how the Christmas turkey must feel, wrapped in silver foil. It was certainly warm, and the atmosphere was added to by the appearance of the *Fisherman's Friend's* singing group who were raising money for charity. Barry, the owner, and his assistant Woodbine (not his real name) were busy serving the many locals who had filled the pub for the singing of Cornish shanties. The heat was building and soon the doors were opened to try and cool the place down. We never did get dinner that night and headed home hungry.

John came up with the brilliant idea of having a motorhome to transport the dogs and cats when we eventually made the epic journey from Cornwall to France. The pets would be with us and happy, so he searched for what was available. Meanwhile, my book was nearing completion. I'd found a publisher and we'd agreed to publish it in September, so this was something of a boost to morale. We travelled to Dordogne again at Easter to view and dismiss houses and have a short change of scenery.

Thankfully, after a few weeks we had another offer. It was slightly less than we would have liked, but the estate agent said a very interesting thing to me.

'Yes, you could hang off for a few months for another possibly better offer, or you can take the loss and go to France and start the rest of your life.'

Life is short and tough. Everything we had gone through with my mother highlighted that life was for living. So we would lose a couple of thousand on the offer, but had gained over a hundred thousand by having the agricultural tie lifted. I could live with that.

I accepted the offer. We now had a date to move out of the

house by mid August and time was catching us up. Soon the contract had been signed and we were packing in earnest. We didn't know a whole lot about the new prospective owners, but after our last experience, this suited me completely.

Searching the internet for suitable French properties showed that the same houses were on with multiple agents; but as it was now spring, new properties were appearing, and one particular one kept coming up. The photographs were misleading. It seemed a strange set up and much bigger than we needed, but the realisation that, as early retirees we needed a business to get us into the healthcare system, which meant that it could offer the possibility of gite rental. So, I arranged to fly over alone for a weekend in May, whilst John sold off even more equipment. I'd arranged to view four properties, and met the agent at his office. The schedule announced that the first property was an immaculate, modern house with an attached annexe, but when we drove up to the property it was clear the photos had been taken some time ago. Immaculate it certainly was not.

The wooden gates were warped and the "lawn" looked like a neglected hay meadow. The agent explained that I needed to "see past this", that the English owners were going through a particularly messy divorce and, as a result, the house had been empty for a while. Inside it was clear there had been a substantial water leak and the house had been neglected. As a veteran of dealing with many renovations, I could indeed see past the cosmetic and was pleased with the location, the manageable amount of land and the layout of the house. It was large, but there was an existing annexe that was already used as a gite, and the possibility of separating off a second smaller gite at the rear of the house. The large swimming pool to the front of the house needed refurbishing, but would be a great attraction for holidaymakers. The price, however, was outside our price range. I walked round one last time and went to see the other properties, which were not what we were looking for. So, I drove back to the first house and walked around it alone, and then the village it was located in. After an hour, I parked up and phoned John.

I sent him my own photos of it and the village and explained the teensy little issue with the price, and said I'd try to secure a second viewing before I flew back the following evening. Sadly, the agent was fully booked with other viewers at other properties. So, I did the only thing I could: send information to my friends in France, asking how much we could negotiate off the price. I knew then that this was the house. Yes, it was considerably more than we had to spend and would need a fair bit of renovation, but it was modern, well insulated and would provide us with a business. A few days and many phone calls later we had agreed a price, a date and our offer was accepted. Kieran was pleased, if a little surprised at the size of the place and asked if we were going to get sheep for the one acre paddock. I firmly told him that we had done with that sort of life. We would not be having any livestock again. We were going to relax now.

Back home, John had bought an elderly but well-looked-after motor-home and was servicing and adjusting the interior layout so we would have room for two dog crates to contain the dogs whilst travelling. We had arranged to hire a storage unit for a month or so and began to look for removal quotes, which was quite eye-watering. Colin and John took loads of stuff to the local tip in the car, and sold even more equipment. How he had managed to store all this stuff was beyond me, and I had to assume the shed was some sort of Tardis that was actually bigger inside than it looked. As our agreed date to move out of the Cornish house was three weeks before we signed and moved into the new French house, we had to arrange storage of our belongings locally, and the huge metal container was soon half-filled. The plan was to take the motor-home and the dogs up to Scotland to see friends and family before we left the UK, and planned a few weeks there to revisit some favourite places. This would be a good trial run for the dogs and ourselves to see how living in a motor home would be. I visited some college staff to say goodbye and on Friday went for a relaxing massage nearby.

I arrived home relaxed and smelling slightly of lavender oil, surprised to find John sitting on the sofa, in clean clothes.

'I had an accident,' he said, grimacing. 'We need to go to the hospital at Bodmin.'

He had climbed the rear ladder of the motor-home to clean the roof and had fallen spectacularly from the top, landing on his left shoulder. Knowing instantly by the pain this caused, he staggered back to the house, crying, managed to strip and have some sort of shower. He had shrugged a shirt on in preparation for an X-ray. So, I drove him, very slowly to Bodmin. The nurse took one look and told us to go to Truro. She couldn't deal with this sort of injury there, and so we drove the thirty minute journey very slowly to the large hospital at Truro with John in excruciating pain.

Trying to be the model wife I said nothing, whilst in my head calling him a total dimwit. *How the hell were we going to manage a house move and a trip to Scotland and then a huge drive to France?*

About three hours later with a diagnosis of both a broken and dislocated shoulder, an injection of various painkillers and a successful manipulation by three burly male nurses and a bed sheet, we were on our way home. Naturally I was upset that he was in pain, but was more worried that I would have to drive the motor-home to Scotland and probably to France. It was huge, an automatic and a left-hand drive. I'd never driven anything like that before. The relaxing massage had been wasted and I was more stressed now than I had been beforehand.

On our final day, our friends rallied round and helped us move the last of our things, and we were only an hour late letting the solicitor know we had left the house. John was dosed up with painkillers and was visibly tired, and sore. I didn't really have time to be sad or reflect on all the things you normally do on leaving a home for the last time. Exhausted and had so many things to think about. The solicitor who had been dealing with the sale was off on holiday and I spoke to the senior partner to ensure everything was okay and that the funds had been transferred to my bank account. As this was a Friday, I would check sometime at a bank cash-point, but he told me it was all done, and wished me luck.

Saying goodbye to all our friends in Cornwall was hard. Saying goodbye to Colin and Dave was hardest. Colin had been a staunch friend through our hardest times and Dave had in many ways replaced my father, who had died when I was in my twenties. We hugged and cried and Colin eventually had to take Dave away. Saying goodbye can be miserable.

We drove off eastwards and stopped just outside Bristol for the night, exhausted and glad to be on our way to Scotland. The next morning I checked the bank to discover that the money had *not* been transferred, and had to wait till Monday morning to contact the solicitor to discover the reason.

*****

I'm sure you can imagine my worry over the weekend, and the ensuing phone call to the solicitor at 9am on Monday. So, to spare you (and me) I will condense the facts.

Apparently, the money had been transferred to an account I had closed some months previously, after the first sale had fallen through. The email I had sent confirming the *new* account to be used for the proceeds, although acknowledged by the solicitor dealing with the sale, had never been passed to their accounts department. They said. So the money was missing. I think I was actually speechless for around two minutes as the horror of the situation manifested itself in my head.

Nearly half a million pounds had simply vanished. The senior partner was "making enquiries", whilst I, at my end, also made enquiries, when I wasn't hysterically crying. Between bouts of frustrated tears I manoeuvred the motor-home along the motorway and headed north. At least concentrating on the road took my mind off the shambolic actions of my solicitors.

The story about the missing money continued, with the solicitor's bank, my bank and the old bank all denying they had the money anywhere. This went on for over a week. I was back on my steroids, panicking that we would not get the money in time to transfer it to France to complete the new purchase. Meanwhile, there was huge volatility in the exchange rate between sterling and the euro, due to Brexit uncertainty and we watched with horror as daily the pound became weaker and weaker. As the pound weakened we would be able to buy less euros, not a huge issue if you were exchanging a small sum, but a huge amount when the amount was hundreds of thousands. This meant that to pay the balance for the French house we were now eating into what was to have been an emergency nest-egg. But we had no option but to wait and hope someone would find the money.

We visited family in Glasgow and headed around favourite spots in Scotland, but the joy of the long anticipated journey was like ashes in my mouth. The dogs had a wonderful time and if all had been going smoothly we would have enjoyed it too, despite the rain. As it was every time we stopped to rest, I was unsuccessfully calling the various banks involved to try and find the money. Typical Scottish summer weather reminded of why we had decided against returning to Scotland and move to more reliable weather in South West France. Oban? I should cocoa.

Finally, we made progress, thanks to a fantastic assistant in one of the banks' main branches in Glasgow. She was like a terrier, refusing to let the incident lie. She discovered where the money was and it was in the process of being returned to the correct bank account, but would take a few more days. This efficient (and ruthless) senior assistant had made it her personal mission to sort this. Eventually, after a total of thirteen days, the money was in my account. She phoned to tell me and, after I had stopped crying, she advised me to make complaints to the Law Society regarding the solicitor's incompetence (she rightly called him a 'numpty') and the Banking Ombudsman for advice for the shoddy and slow response from two of the banks involved in the fiasco.

Once I assured myself the money was actually in the account, we began to drive South. On the Monday I phoned the delightful man at our currency transfer service and discussed the poor exchange rate and our funds. He fully understood our situation, but due to the plummeting pound, we could exchange enough to pay the balance of the house purchase, but didn't have enough to pay for the removal firm or fuel from the ferry port in France down to Dordogne. Due to the thirteen days the money had been missing we had lost nearly thirteen thousand pounds! Thirteen turned out to be very unlucky for us. We reluctantly arranged to borrow money from family, and with heavy hearts we drove back to Cornwall where friends with a campsite offered us a free place to park for the few days remaining till we left the UK.

It was a long drive from the French ferry port to Dordogne, stopping every few hours for the dogs and ourselves. We finally arrived late on the Sunday afternoon. On the Monday we attended the offices of the notaire, signed for the house and finally had the keys. The removal van, mainly containing equipment, tools and books would not arrive till the following week, so we moved into the gite which we had thankfully bought fully furnished. Just as well really, as we had no savings and very little furniture to go into the main house.

I made a complaint to the Law Society which was partly upheld, and to the Banking Ombudsman, who were horrified and extremely helpful and supportive. The case officer in charge was exemplary, phoning or emailing me at least weekly to reassure us that they were making progress and nailing every lie that one of the banks had told me in an effort to cover up their mistakes.

One day in early December, whilst in the garden, there was a text message on my phone asking me to phone them as a matter of urgency. I hurried inside, with apprehension at the urgent demand. Surely, there wouldn't be more bad news? When I got through to her on the phone, I could hear the note of excitement in her voice. Slowly and clearly, she told me that a sum of just over £13,000 had been transferred to my account in the last hour. I was so stunned I asked her to repeat this. It had taken four months but they managed to recover over £13,000 that we lost.

The build up to Christmas was a real celebration. New friends helped us secure firewood, chimneysweeps and with various day-to-day things, and our first Christmas in our new home was quiet but happy. We discovered acquaintances near Bergerac and in the Lot. Our French neighbours waved and called 'Cou cou!' when they saw us working in the garden, and we were welcomed at the local bar and in the village. They were delighted that we intend to stay permanently rather than use the house as a holiday home and have welcome us into the community.

By March, I ended one fine, warm, blue-sky afternoon

sitting in my now orderly garden by making my weekly phone call to Kieran, and casually mentioned that the next day we were collecting a couple of lambs for the paddock. He laughed and I had to laugh too. I confessed that I had missed having the sheep so much that the arrival of spring has been too good an opportunity to miss. You can take the person out of the farm, but you can't take farming out of the person. A smallholder I was, and still am. I'm just living the Good Life in France now.

Printed in Great Britain
by Amazon